Rich App Poor App

Simon K Williams

www.AppManSecrets.com

Published by ThinkEmotion
Design by ThinkEmotion

14 De Beauvoir Place
Tottenham Road
Islington
London
N1 4EP
England

First Edition

ISBN: 978-0-9569250-0-8

Dedicated...

To all those that look forward not back, take action not stagnate, dream not ponder, strive not give-in, live not exist, laugh not moan, embrace not reject, support not ridicule, teach not preach, grow not recede, yes not no...

To my mother Katheryn who was taken too soon, for me to give back...

To Graham my father, whom I cherish and love...

To my girlfriend Adele, my support and my beacon of light...

This is for you

iPhone and iPad applications offer a global commercial opportunity that no other mechanism can provide.

With application sales measured in their 10's of Billions, there is little surprise you have decided to investigate further, and well you should...

Like any industry there are *Best practice* techniques you should follow, and a number of easy-to-avoid (if only someone told me) commercial traps, that you can't afford to fall into.

This book will take care of you throughout your entire journey, from a blank page, to an App Store hero.

Your exciting journey starts right here...

What others think...

"At last, there's a book crammed with insider secrets that show you exactly how to optimise the new world of 'App-ortunity'. User-friendly and with compelling clarity, "RichAppPoorApp" sets out the blueprint for anyone seeking to take a concept and make it a commercial reality. This book is thoroughly engaging and encouraging in equal measure, offering a glimpse into a future that's already happening."

Mike Symes
Managing Director, Strand Financial

"Rich App Poor App, is by far and away the most useful and practical guide for anyone embarking upon their own App. If you follow the development structure that Simon outlines you'll have a high-quality product that will both make you money and give you considerable Kudos. In my experience as soon as people know you have an App in the App Store you are held in even higher regard than those with only books and CDs. Buy it, read it, DO IT!"

Sam Brown
Author of several Apps, including Drum Secrets

"Apps are a multi billion dollar industry already but in the future a business without an App will be like a business without a website. Clearly apps are an ever growing opportunity. Simon has spelt out the clear opportunities in this market and given the reader a framework to follow. Add in your own unique take on it and your well on your way to new income and/or engagement with your market"

Daniel Priestley
Author, Key Person of Influence

Contents

Contents

Foreword

When I was at Mercury Communications in 1993 we dreamt up a device we called a Smart Palm. It came out of our mission to deliver People, Information and Entertainment into the palm of peoples' hands.

In prototype it was remarkably similar to today's smart phones and tablets. We worked with both Nokia and the Apple Newton project to see if we could get something real to work, but we were well ahead of our time and the available battery, display and communications technologies were just not up to it.

I had to wait for a new millennium to see that old Mercury dream come to fruition and it was of course Apple who brought it to life with the iPod, iPhone and iPad.

Whilst the development and evolution of the hardware was foreseeable, the emergence of a new way of providing, and engaging in digital services was not. In June 2008 the App concept was born, and with it a shift in consumer behaviour.

Consumers now have the ability to acquire highly specific, modular, bite-sized solutions on almost any subject matter, delivered straight to their pocket through a near instant digital distribution method. The ripple effect of these devices has far reaching consequences, impacting every industry across the globe, which is why it is essential to get the facts on this emerging communication method.

As the Chairman of Garlik, an award winning technology company, which helps people take control of their personal information and protect themselves against identity theft and financial fraud, I felt it was important we explored this exciting new platform further. I would urge you to do the same.

Fortunately I was able to get my hands on an early copy of Rich App Poor App. I liked the easy-to-follow process, the sound business strategy, and the highly revealing practical exercises.

Since then I have been engaging very closely with Simon and I really value his insight. His insider perspective, inspirational guidance, and pragmatic approach will ensure you avoid the pitfalls he speaks of in this book, and enable you to maximise your personal, and very unique App opportunity.

Mike Harris
Founding CEO of Firstdirect and Egg plc
Co-founder and Executive Chairman of Garlik
CEO of Mercury Communications 1991-1994

About the author...

Simon K Williams started his career as a Graphic Designer working with house-hold names such as EasyJet, Sony, EMI, Hilton, Thistle Hotels, The Labour Party and Virgin to name but a few.

At one time or another he has designed and produced about every form of printed media there is, from small newspaper adverts, to CD cover, to road-side billboards.

However, it was his time working at a Contract Publisher that allowed Simon to make the transition from making products look beautiful, to products that make money, and thus quickly progressed from Graphic Designer to Assistant Publisher in less than 18 months, securing significant commercial contracts along the way.

In hind-sight Simon was evolving from a Designer into an Entrepreneur, or as he calls it 'a designer with balls'.

He expanded his expert field of knowledge by completing and excelling as a Masters student in Product and Service Design at Middlesex University, where he now advises current students on design and innovation opportunity as a senior lecturer.

The move away from the immediacy of Graphic Design into Product and Service design showed Simon that due to the creation, production, manufacture process, all designs must be created for the future. If you only design for today, it will be out-of-date by the time it hits the shops.

This new fascination for the future led to the creation of his first book *Make Your Product Invisible – or nobody will buy it*, which uses sound scientific research to suggest that the product and services that western society will desire will be based on producing and providing emotionally rewarding experiences, more than delivering practical benefits.

That book reveals cutting edge product creation techniques which enable designers, entrepreneurs and business owners to produce the products and services our society craves, a few of which will be revealed

Simon K Williams speaks around the world from Singapore to London, delivering valuable insights, so that App creators everywhere are able to maximise their commercial potential on their very first attempt

within these pages. It is as he says, his finest work, and has led to countless exciting opportunities, including several business partnerships.

Perhaps it's no surprise that he was quicker than most to recognize the commercial potential of the App Store several years ago, and acted on it swiftly, producing the first ever Metro navigation App for Moscow.

Four months later he was generating enough passive income to quit his full-time position using the royalties to achieve one if his primary goals by starting his own very own App development Company.

At the time of writing, Simon and his development team have produced Apps for four different Countries, over 30 different industries, with over 100 applications for the App Store, and has provided guidance, insights and direction to thousands of budding entrepreneurs, business owners, and the simply curious across the world.

All the Secrets of the industry, all the knowledge, and all the techniques are revealed within these pages to ensure your first App is as commercially successful as another developer's 100th.

Avoid rookie mistakes

Thousands of new and exciting Apps are entering the App Store all the time, the disappointment for me is seeing so many that have been constructed in a way that prevents them from maximizing their potential, and in some cases all but prevents them from making any sensible commercial revenue.

Mobile applications are an exciting emerging territory, which is attracting lots of newcomers for all the right reasons. However, because of this, there is a great deal of ignorance surrounding the industry, ignorance caused primarily by a lack of experience. As a result many Rookies, dare I say it *most*, will never produce a commercially successful application on their first attempt.

This book's goal therefore is to give you the tools, the insights, and the industry insider secrets you need to ensure that your very first application is a commercial masterpiece. Learn from the mistakes of others so you don't fall into the same trap that haunts most App Rookies. Here are just a few of the mistakes you will learn to avoid...

- ✗ No expert advice
- ✗ No concept protection
- ✗ No competition assessment
- ✗ No concept testing
- ✗ Focus on personal desires
- ✗ No user research
- ✗ Make bad assumptions
- ✗ Consider writing code themselves
- ✗ Limit value in App
- ✗ Rush in without advise
- ✗ Use heart not head

- ✗ No App Marketing (not even free methods)
- ✗ No feedback loops
- ✗ No way to capture users data
- ✗ No strategy to ensure positive user reviews
- ✗ No product funnel
- ✗ No App evolution
- ✗ No focus on the objective
- ✗ Complete a formal Business Plan
- ✗ No action taken, let others take their App opportunity away

Rookies

Many Rookies have half a descent idea, assume it is going to change the planet, doesn't require any knowledge, and will earn them millions over-night because they read something about it in the paper that morning. In my experience it doesn't work like that.

There are sound proven principals to creating commercially successful products and services. This book adapts traditional best-practise principals to fit the exciting emerging world of iPhone and iPad applications. Most of the suggestions in this book are simply common-knowledge to anybody working within the industry, however they aren't common-knowledge with App Rookies. The words in this book will address that, giving you all the tricks usually only available to seasoned App developers.

In-fact, since most App developers have no commercial interest in how successful your application is in commercial terms, they will simply follow your instructions, make your App look pretty (if you're lucky), and run. They aren't behaving irresponsibly, the truth is they have no idea of what makes an App sell, their knowledge and experience is in software creation, not commercial business.

By the time you reach the end of this book, having completed the easy-to-follow exercises along the way, you will have generated 15 App concepts based on your own interest and knowledge, assessed there commercial viability using sound statistical information freely available, identified your unique App-ortunity, and learned exactly how you can bring that concept to the App Store by using best-of-practice App marketing principals – most of which are available to you for *free*.

You will have learned about Blast marketing techniques, which will ensure that every ounce of effort is used wisely, so that you maximize the impact your application has when it lands with a thud, in the App Store.

The very first exercise I will request of you is goal related. It may surprise you, that many App creators have never actually defined what their App goals are, as a result it is very difficult to achieve them. So first-things-first, what are your App goals...

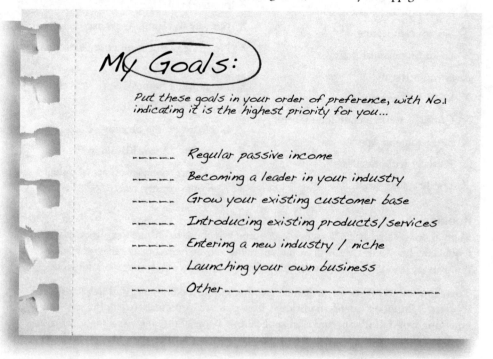

My Goals:

Put these goals in your order of preference, with No.1 indicating it is the highest priority for you...

_____ Regular passive income

_____ Becoming a leader in your industry

_____ Grow your existing customer base

_____ Introducing existing products/services

_____ Entering a new industry / niche

_____ Launching your own business

_____ Other_____

What can the iPhone and iPad do?

There are at present three main devices that can be used to use Apps from the App Store, the iPhone in it's various forms, the beautiful iPad, and the almost entirely forgotten about iPodTouch. Since the iPhone is the most feature rich device, and the device most frequently used for App downloads, we will take a brief look at it's technological features.

FaceTime: Arrived for the first time in the iPhone 4 which made it's debut in June 2010. The iPhone contains two cameras the main one, on the back of the phone as you expect, and another much smaller camera above the screen facing the holder, this provides the opportunity for video-conferencing.

FaceTime allows face-to-face communication between any two iPhone 4 users, as long as they are in a good WiFi location. This provides the bandwidth to allow video-streaming to work. This type of technology has been promised for many years, however this is the first consumer level device to bring this technology to the masses.

As Apple, puts it *"With just a tap, you can wave to your kids, share a smile from across the globe"*.

Retina display: The screen resolution of todays iPhone's are impressive with four times the resolution of previous iPhone's, the pixel density is so intense that the pixels can not be detected by the naked eye, which results in a perfectly clear and beautiful image.

Whilst display improvements don't lead to any new features as such, this does mean that every experience is heightened and improved as a result. This is particularly useful for video playback, which is a growing area of iPhone use, and in line with Apple's own plans to dominate the movie-streaming business.

Multitasking: This was something that many owners of older iPhone's requested, and actually an area that Apple's competitors got to first. However although Apple, didn't get their first, they may argue, they did it best. So users can now jump from one application to another without needing to close the previous one down. This leads to faster jumps between your favourite Apps, but also allows more integrating between the Apps, as one App can access another for data potentially.

HD video recording and editing: With the improved camera now available, you can shoot in HD video straight from your iPhone, which can work effectively even in low-light situations. As if that wasn't enough, you can use Apple's own *iMovie* App to edit, add titles, transitions to create fantastic short video's on the go.

5 megapixel camera with LED flash: Improved resolution improves every photo, and every video clip you shoot. The LED allows good shots to be taken even in low-light situations. Although the forward facing camera is primarily used for FaceTime, it can also be used for self-portraits.

iBooks: Is an extension to the App Store, providing digital copies of thousands of available books, whilst it is possible to view these on the iPhone, they really come to life on the iPad, with beautiful turn-page animation, the ability to bookmark pages, and look up words in a dictionary. These are very digitally elegant, offering an immediate pay, download and read option, perfect for anyone who is in a hurry for a particular book, or essential reference.

Other features include; folders for App collections, custom home screen, iMovie, Mail, Safari, iPod, Voice control, Maps & Compass, Keyboard, App Store, iTunes Store, Accessibility, Voice memos, Calendar, Stocks, Nike + iPod, Weather, Notes, YouTube, Calculator, Parental Controls, Encryption, Tethering, find your iPhone, and more

More than technology

As we have seen, they perform a host of other tasks outside just telephone calls. It's easy to look at the technology, but what is more important is to look at the capabilities, which is very different...

It knows who you are

Knows your name your age
Star sign
What music you listen to, what films you watch
How much you spend on music / films per month
It knows where you live
Where you like to go on holiday
It knows how much you walk, take transport

It can talk to you

It can read to you
It can play you music
It can answer your questions
It can tell you how well you are doing something
It can give you feedback

It can show you stuff

Can display your favourite photographs
Display art
Show your video clips, or home movies
Can show you clips from other people
Show you a feature film
Show you a cartoon, book, newspaper, magazine
News clips, sporting clips,
Graphs, tables, spreadsheets,
Display messages from friends, colleagues, business

It can listen to you

Listen to what you say, and perform what you ask it to do
Can pass on your message to anybody you have the contact details for
Can record what you say, and play it back to you when you want

It can touch you

It can vibrate when somebody special calls

When you perform a goal, request, or challenge

Give you vibration feedback when you touch a button

It knows where you are

It can tell you whats nearby

It can tell you how to get to your next location

It can show you how far away you are from friends, buildings, services

It can give you directions

It knows how high you are, altitude

It knows if you are in sun, or in rain

It knows which way you are facing

It can tell you what you are looking at (*EyeView* App)

It can tell you who you are looking at, and details about them

It knows how fast you are going

It can tell you how fast you are running, cycling, skiing, sailing, driving...

It knows what angle you are holding the device

It knows if what angle it is at

Merge all these laments together, and iPhone's and iPad's know almost everything about you – the opportunities are bewildering, endless, and exciting.

Just think about it, it knows your name, where you live, how old you are, where you are, if it's a warm day or not, how high up you are, what you are looking at, who you are looking at, how you got there (did you walk, or take transport), and it probably knows what you are doing next (calendar). The power of these devices are astonishing – and the truth is we haven't even scratched the surface of what is possible.

Can you think of something cool it can do?

Consumer behaviour

To fully understand your commercial opportunity, you must first understand the significant move in consumer desire and behaviour that is occurring across the Western societies. This has powerful consequences for any person, or organisation wishing to produce and sell commercial products. Understanding these social factors will allow you to create products of sustained high demand. The types of products that any individual or Company would be proud to produce.

I investigated this exciting cultural shift in my first book *Make Your Product Invisible - or nobody will buy it*, released before the App Store was even launched. So I released the significance of the App Store as soon as it was revealed.

Customers require products and services that connect with them in a highly personal and individual way, appealing to them at a human, rather than consumer level. This is not going to be a passing fad, but a significant development and direction in consumer behaviour. Companies that embrace this phenomenon will be highly successful, and those that don't will find their services swiftly losing sales.

And this prediction has proved to be the case.

Emotional products...

As Rolf Jenson explained in his book *The Dream Society*, consumers are searching for Emotionally products, and to achieve that the consumer must feel they have a relationship with the product.

This explains why the most popular products of recent years have been significantly different to those that proceeded them. Phycological products such as TV's Big Brother and X-factor are now reporting the highest TV viewing audiences, because their the entertainment they provide is emotionally focussed.

If you have not witnessed the X-Factor phenomenon, then it is a music contest series where singers and bands audition and then compete for the ultimate prize a multi-million pound recording contract with one of the industries leading promoters Simon Cowell.

Early episodes focus on bad auditions as much if not more than the good, with every episode recording the highs and the lows of individual contestant stories as they attend to win the ultimate prize.

As the series continues, the audience gets more and more connected to their favourite artists, spending their own money to vote their preferred performers through to the future rounds.

The show takes a deep look into the emotional journeys of it's contestants, and makes no serious attempt to find the next Elvis, Beatles, Madonna, or Nirvana. The more stressful the situation they make it for the contestants, the more enthralled the audience is in witnessing their often traumatic journey.

There are many ways to make an experience emotional, one of the very best ways is to connect to the audience at a highly personal level, examining their unique desires, needs, passions and wants.

Companies or individuals that deliver these emotional experiences will be highly successful.

The long tail market

Amazon are the biggest and most famous book selling service on the planet. A large book shop can only hope to stock a maximum of 140,000 books, Amazon sell more than that in an hour, and has a list of products that go into the millions, but here's the amazing bit...

Amazon sell more books outside their top 140,000 titles, than within it. This is a clear example of what's known as the long tail phenomenon. So whilst there are a few block-busters selling millions of copies, the vast majority of us are searching, finding, and most importantly buying highly-niche products, which speak to us about a highly specific need or desire.

> **❝ Amazon sell more books outside their top 140,000 titles, than within it ❞**

It has often been said, that we are what we wear, and to a degree that is true. It is clear that what we wear, use and acquire describes who we are as people, which is why so many of us care deeply about the cars we are seen to drive, the holidays we take, and the homes we live in.

Generally speaking we all like to be seen as highly unique individuals, so what better way to demonstrate that uniqueness by purchasing highly specific niche products and services such as the highly original books that only the virtual book stores such as Amazon can provide.

It has proven to be a successful business strategy. Amazon reported a 68% rise in net profits to $299m in the first quarter of 2010, that is a meteoric growth, with total sales of $7.13 billion.

Apple have taken this proven formula, and brought it to the mobile market. Selling music, PodCasts, iBooks and Apps, via iTunes – and it's worked on a huge scale.

As a result, Apple are also seeing startling growth figures, with a revenue rise of almost 50 per cent in early 2010 compared to the same quarter of the previous year, profits were up 90 per cent. These results do not even include iPad sales which are expected to add another $1 billion (£654m) of revenue (Times).

Overall, the Company reported net income of $3.07 billion, or $3.33 a share, up from $1.62 billion, or $1.79 a share a year before. Revenue was $13.5 billion compared to $9.08 billion (Times).

As a result Apple have overtaken Microsoft to become the number one technological Company in the world. It seems this strategy has worked rather well for Apple too!

Repercussions of the long tail market

The desire for highly niche products suggests a significant trend away from huge conglomerates, and the one size-fits-all approach

Our future will contain less high street chains, less face-less franchises and more family run niche businesses.

The power will be returned back to individuals, the common man (and women) will have greater power to deliver products and services to all, the internet has been a major tool to help this, but I want to make this clear...

Society is NOT moving in this direction because of the technology, it's the other way around. The technology was created, and adopted because of the needs and desires of society, the technology just helps to achieve it.

Why do Apple only sell one mobile phone, and does this fly in the face of the Long-tail phenomenon?

Prior to the release of the iPhone every significant mobile phone manufacturer offered a wide range of products, in an effort to meet this need for uniqueness.

However the users product experience, includes many elements other than just the physical product or device, including marketing, tone-of-voice, shop design, shop staff, customer service, and post-sales service to name just a few.

> **❝❞ it is important to note that pleasure with products accrues from the relationship between a person and a product. Pleasurability, then, is not simply a property of a product but of the interaction between a product and a person ❞❞**
> Jordan 2000

Recognising this, Apple have provided the emotional experience by providing access to thousands of applications that will cater for almost any desire, as a result there was not the need for a huge range of handsets.

This enabled Apple to place a huge amount of technical functionality inside the one handset they did offer, including revolutionary features such as an accelerometers, compass, and a gyro, which would not have been feasible if they had to create a long line of phone models.

Since the iPhone shook up the industry, all rival manufacturers have drastically reduced their handset range to a few select models, but without a similar ability to personalise the phone via mobile applications the iPhone has few significant challengers in the market place.

World-wide audience...

If you're in the business of selling micro-niche or long-tail products, then a wide distribution method is essential. Both Amazon and Apple rely on the Internet to distribute their products. This business model simply would not be visible by any other current distribution channels.

It's this need for a wide distribution channel that explains why Apple has been so determined to establish iPhone sales partners in so many of the worlds Countries.

This is phenomenal achievement, which has often been under appreciated. At the time of writing, the iPhone is on sale in 94 Countries around the world, including Countries such as Russia and China which are traditionally challenging territories to reach agreement with.

Should you invest in Apple?

I'm not going to attempt to assess every financial vehicle available, but lets have a brief look at the biggest investment options;

> Bank savings
> Stock and Shares
> Pensions

Banks

In the worldwide recession that hit in 2007, it was the Banks that got hit first, and hardest. Some collapsed, some were bailed out by their governments at their Tax payers expense. This huge industry was hit so big, and so hard that it has been irreversibly changed forever.

140 US Banks collapsed in 2009, with even more Banks collapsing in 2010, few are handing out loans, mortgages, or attractive Savings accounts. Instead savers and investors are forced to look for alternative opportunities.

Stocks and shares...

Equally hard hit, were International trade markets, billions was wiped of the value of stocks and shares. With huge Companies like Barings and Enron collapsing, British Petroleum (BPI) and BA Airlines losing billions, few people are trading – very simply most are too scared to invest.

In traditionally difficult trading times, investors would run to the relative safety of Utilities, Gold and Banks, however these stocks are either over-valued or unstable. Pensions are no better, the UK has a significant Pension deficit and simply will not be able to afford to deliver the Pensions it had once promised.

So for most, stocks and shares are not a tempting investment.

Property...

Property is seen as an investment in some Countries more than others. In America and the UK in particular it is seen as the key industry. When property prices reduce the whole economy suffers.

In the UK there is a massive housing shortage and property was seen by most as a secure robust investment. Whilst the demand for housing hasn't diminished the ability to obtain mortgages certainly has, as a result the property industry has been devastated, with many owners going into or fighting negative equity, and first time-buyers unable to get the mortgages they need to move.

The lettings market has survived a little better, although rates are generally down. Any property recovery is likely to be slow at best, with fears of the so called double-dip, meaning there are real concerns of another property slump, so understandably people are again too scared to invest.

Doodads...

Robert Kiyosaki taught us all in *Rich Dad, Poor Dad*, to invest in Golden Eggs, not Doodads – Doodads are what Robert called products of any kind that do not offer any return on investment. So a new TV, iPod, is considered a Doodad. A doodad, that requires ongoing maintenance or investment such as a car is considered a liability. Robert taught us instead to invest in financial vehicles that can offer a return, but as we have seen the traditional vehicles aren't worth investing in.

So what now...

Rather than looking for others to make money for us, we can take control ourselves. By investing in ourselves we can create wealth, and the potential to magnify our investment 10s, 100s, 1000s of times over. This is the approach I practice and recommend. After all, if you are not willing to invest in yourself, why would anybody else?

The App Store provides a unique opportunity for you to introduce your products and services to a world-wide audience, an audience that has the ability to purchase your goods 24 hours a day, 365 days a year, from the office, from the home, from the bus. There has never been a device capable of selling so much, and in so many situations.

Is Apple safe?

Many global businesses collapsed in 2009 and 2010, so could Apple be next?

The Banks business is based on our confidence in the stock market. In effect it's a gambling game, and rests it's chances of success on the fact that over time, wealth of the worlds largest Companies will increase.

The long-tail phenomenon, suggests this may not necessarily be the case, as people turn towards smaller Companies for highly niche services.

Apple rests it's success on Consumers paying good money primarily for these products, it delivers an environment so that almost anybody in the world can buy a highly niche, Song, TV show, PodCast, Audio book, Digital book, or App.

This is why Apple is registering it's highest ever quarterly profits, in the same months as the Banks were collapsing. Apple's products have never been cheap, quite to the contrary they are considered luxury, prestigious desirable products, and despite tight-wallets, people are finding the money to invest in the ability to own, use, and enjoy these highly niche products.

Products describe who we are, and we all want to be seen as an individual, that's why these products work, and why Apple are safe, and why I choose to invest in them.

Apple's profits have risen sharply, at the same time as other institutions have experienced dramatic fall in profits, leading to the collapse of many of the world's banks.

What would have the growth of Apple been like, in a more positive financial environment?

Apple have looked at the Amazon business model, brought it to the mobile audience, and arguably done it better.

Whilst consumers usually focus on the physical products, such as their beautiful iMac, iPod, iPhone, and iPad in Apple's case. Steve Jobs himself said that Apple is a software Company, they just house that software in a nice box, because it's the software that their customers are engaging with.

Apple provides both the financial size, and the future prospects and strategic positioning to offer an exciting opportunity, and an investment opportunity either as a share-holder or as App developer that few, and in my opinion that no other Company can offer.

Smart phones

From around 2005 the devices moved away from being purely telephone devices, to personal aids, and although they are still referred to as phones (just as CD shops are still often referred to, as Record shops), for many the core services they delivered was a contacts, diary, lifestyle assistance, of which making phone calls was just one element.

66 By 2012 there will be more Smart Phones sold than traditional phones worldwide 99

As the technology included expands, the lifestyle services it provides also expands. There is clearly a desire for these life-style products as reflected in the surge of 'Smart phones' against the more traditional telephone biased alternatives. Symbian, RIM, iPhone, Windows Mobile, Linux, Android all report that their Smart phones are the largest growing sector, with the iPhone selling nearly ten times as much in 2009, as it did in 2007. By 2012 there will be more Smart Phones sold than traditional phones worldwide.

So the trend is clear, people do indeed want lifestyle aids, not telephone devices. Traditional mobile Companies were slow to adopt the smart phone approach fully, their hesitation, and lack of commitment was the open door that Apple needed to dominate.

Companies such as Sony Ericsson, Nokia, Samsung have all been unable to react quickly enough. Indeed even Google has been able to enter the market, following Apple's lead, as the industry is still too slow to react.

In 2011 the only App Stores available other than Apple's is the BlackBerry App World, and the Andriod Market. These are gaining traction, as some developers wish to claim their stake of land on these other platforms, but Apple is the still the dominant force for now, and the foreseeable future.

By 2012 there will be more Smart Phones sold, than traditional phones worldwide

Unique App-ortunity

- ✓ 120+ million App Store users
- ✓ Who spend $4.37 on Apps every month
- ✓ On 213,979 Apps available
- ✓ Made by over 30,000 developers
- ✓ Which make over $300 million in revenue per month
- ✓ 30% is kept by Apple, the rest, is Yours!

iPhone sales per quarter

As the graph below shows the, growth in iPhone sales world-wide has been nothing short of phenomenal, and no signs of this growth diminishing any time soon.

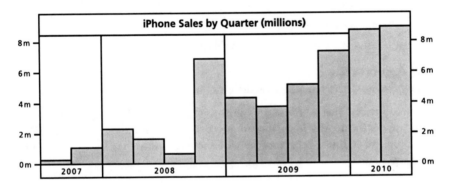

App sales per quarter

Again we can see exponential growth, as this graph illustrates. Please note that this growth does not account for iPad applications which would boost this figure significantly.

The iPad

Many so-called experts in the industry were predicting an Apple flop with the iPad, unconvinced that there was a need for this mini-laptop. However the sales have quietened the critics, with over 300,000 units sold on the first day alone, generating over 1 million in iPad Apps sales in one day.

At the time of launch there were approximately just 100 applications available for the iPad, which is of course growing quickly, yet it doesn't take a mathematician to see the commercial potential in this device.

iPad applications often command a higher sales price than their iPhone equivalents at times, this is because the iPad versions contain further enhancements and extended features, however the key reason for the consumers being prepared to pay more is due to the larger screen size.

If Apple ever bring Apps to the Apple TV (you heard it here first), I suspect we will see another price bump, because again customers will receive a greater emotional experience due to the larger screen size, which in-turn provides an enhanced emotional experience.

The App-ortunity summary

✓ 120+ millions users who can buy your products and services 24 hours a day, 365 days a year

✓ Distribution through 94 Countries, and still growing

✓ Growing demand, with 4 billion applications already downloaded

✓ Fantastic route-to-market

✓ Low entry cost

✓ No expertise or skill needed

✓ It's easy, you don't even need an iPhone

Your App-ortunity

Clearly the App Store offers an incredible commercial opportunity.

When you realise that a billion $ Company, which is also the biggest technology Company in the world, is willing to work with you for free, and to distribute your products and services across the globe to everybody they know, it's hard not to get excited.

> **❝ When you realise that a billion $ Company, which is also the biggest technology Company in the world is willing to work with you for free, and to distribute your products and services across the globe to everybody they know, it's hard not to get excited. ❞**

The exercises that follow will enable you to find your unique App-ortunity. You can get your Rich App Poor App book bonus material, which includes exercise work sheets to help you. Simply visit www.AppManSecrets.com/RichAppPoorApp.html;

App-ortunity: Exercise #1

This is a really simple exercise...

Interest

List 15 areas of interest & knowledge

These can be fields you have studied, areas you deal with at work, topics you read about, or hobbies you pursue. Spend no more than 2 mins on this task, the quicker it's completed, the better –

So get your pen and go for it right now...

- ○ --
- ○ --
- ○ --
- ○ --
- ○ --
- ○ --
- ○ --
- ○ --
- ○ --
- ○ --
- ○ --
- ○ --
- ○ --
- ○ --

Interest *continued...*

List 5 more...

The point of this is to stretch yourself. When you get stuck, that is the point you need to search deeper. Attempt to 'push' through two blank/difficult spots, aim to write 5 more if you can...

- ● --
- ● --
- ● --
- ● --
- ● --

Highlight 5

Five isn't an exact number, if you absolutely possible have to six or seven, then that's fine.

The point of this is to highlight several areas that are worthy of further investigation. However, these are often NOT the same as the first five ideas we think of!

So go on, give your favourites a circle, or a star... ✳

Content

Select one at random

I will ask you to repeat this task with all five higlighted areas.

It's important you don't start with what you consider to be your favourite, because in all likelyhood, you won't then repeat the exercise with the other four.

Venture capatalists have a saying, "I've earned a lot more money from Plan B, than from Plan A!" - it means that the first idea you think of, isn't neccesarily the best.

You can download this, along with lots of other bonus files at **www.AppManSecrets.com/RichAppPoorApp.html**

○ ....

Content media

Now, write down 10 areas of content you have, or can get for that area. These could include; Video, pictures, quotes, Interviews, Case-studies, Games, Statistics, Graphs, Illustrations, Exercises, Surveys, Books, Comics, Articles, Historical evidence, the list could go on and on.

○ ....

○ ....

○ ....

○ ....

○ ....

○ ....

○ ....

○ ....

○ ....

○ ....

Your App-ortunity

Select one at random

I will ask you to repeat this task with all five higlighted areas, it's important you don't start with what you consider to be your favourite, because in all likelyhood, you won't then repeat the exercise with the other four.

Venture capatalists have a saying," I've earned a lot more money from Plan B, than from Plan A!" – it means that the first idea you think of, isn't neccesarily the best.

Feel free to photocopy this sheet, or download it from <web site address>...

- ——————————————————————————————

Content media

Now, write down 10 areas of content you have, or can get for that area. These could include; Video, pictures, quotes, Interviews, Case-studies, Games, Statistics, Graphs, Illustrations, Exercises, Surveys, Books, Comics, Articles, Historical evidence, the list could go on and on.

- ——————————————————————————————
- ——————————————————————————————
- ——————————————————————————————
- ——————————————————————————————

- ——————————————————————————————
- ——————————————————————————————
- ——————————————————————————————
- ——————————————————————————————
- ——————————————————————————————

App-ortunity: Exercise #9

Your App-ortunity

This is where it all comes together, here's an example to give you the idea, and show you how easy this is;

Area of interest: Skiing | *Area of content:* Video

App-ortunity: Ski-Fitness, How to Ski off-piste, Select my best resort, The latest Ski equipment, Mountain side medical advice, How to Avoid an Avalanche, Ski techniques, etc.

It's super-easy, now it's your turn...

Select area of interest

○ _____

Select area of content

○ _____

Your App-ortunity

Come up with at least 6 ideas, more if you like...

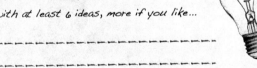

○ _____

○ _____

○ _____

○ _____

○ _____

○ _____

Remember to go through the same process with all 5 ideas - it won't take long, and you will have an abundance of opportunites. I will show you later how to find which is them is best App-ortunity for you

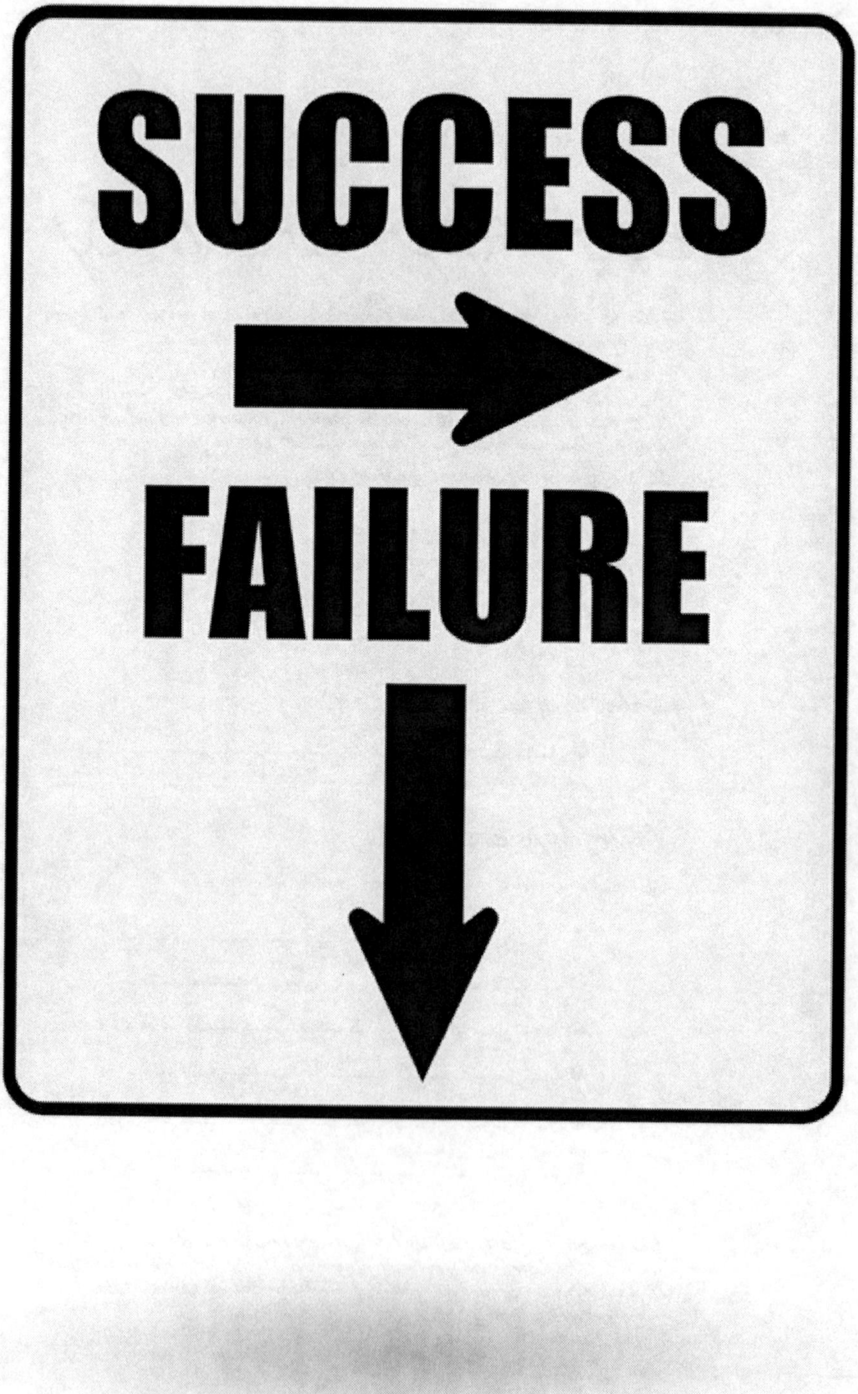
SUCCESS
FAILURE

Know your niche

If you have completed the exercises on the previous pages, then your App-ortunity will be developing nicely, and already have six great concepts. Rookies think of one idea and dive in, pro's think of several great ideas and research to find the best, and then move forward and take action.

Ideo, who are arguably the best Design Consultant Company in the world due in large part to their amazing research techniques, say the way to a great idea is to have lots of ideas. That's why we got you to go through the exercise and squeeze out several great ideas, not just one.

If you have decided to skip the previous exercises, I would encourage you to flick back a few pages. They take very little time, and often lead to amazing results. Even if you are reading this with a firm concept in mind, it is a valuable use of your time.

Idea vs. opportunities

Many new businesses are created on the back of what they firmly believe is a great 'idea'. So why is it that so many of these start-up Companies don't make it into their second year, how can they get it so wrong? Or more importantly, how can you avoid making the same mistake.

I am by trade a designer, who are naturally lazy people! If it weren't for lazy designers we would never have invented the motorcar to avoid walking or riding horseback everywhere. The TV remote control would never have come into existence just to save us the trouble of walking over to the TV to change channel. We have light-switches that are 'clap' operated, digital reminders, and escalators in fitness clubs, so that people who supposedly want to get fit don't have to walk up the stairs.

So as a designer I wanted to learn to take the easy path to success and avoid the pitfalls of those that went before me. After some investigation, I formed a conclusion which I'm happy to share: ideas are worthless!

❝ ideas are worthless ❞

Ask a hair stylist for a great idea and they may have a new hairstyle they think is going to revolutionise the industry. Ask a builder and they may have a cunning plan how to lay bricks quicker. Ask a florist and they may have a new bouquet concept aimed at the younger generation with different arrangements designed to appeal to the youth age group.

The ideas and concepts are inherently based on their existing skill set, or to put it another way, their imagination is limited by their current knowledge. The Oxford

English Dictionary defines *idea* as, *vague belief or fancy*. Would you like to base your business around a vague idea or belief?

To use the florist as an example, it doesn't matter how impressive their new bouquet is, that is going to be a very hard business or product to promote and turn into a commercial success. Whilst it may be a good idea, the commercial opportunity is highly limited, with no obvious route-to-market.

Rather than base a business or product on an idea, base it on an opportunity. Let me illustrate the point...

From time-to-time, I am asked to speak at various events, such as weekend and day long seminars. In many case the audience has paid a significant amount of money to attend, and certainly invested in transport, sometimes flights, nights in accommodation, and at the very least valuable time away from their friends and family. These events can be demanding, and at times the audience members will get tired, thirsty and/or hungry. This is an opportunity. By simply positioning a quiet self-service machine at the back of the room, audience members would be able to refresh and re-hydrate themselves, without the risk of missing a word.

The ability to get revenue from an opportunity compared to an idea is massive. It would be almost impossible to make income from the idea in this example without huge investment in advertising and marketing promotions. By contrast the income of the self-service machines is almost guaranteed. So, when looking for commercial solutions, look for opportunities rather than ideas and you will be on the right track.

66 when I first learned that a billion dollar Company was willing to do a joint venture with me... I got excited 99

When I first learned that a billion dollar Company was willing to do a joint venture with me, distribute and sell my products around the world, and to do all this for free – I got excited. It is to date, the best opportunity I have ever seen, which is why it's now my industry.

So how can we test your App-ortunities commercial prospects, and how can we do so before investing heavily in it's development? We explore some advanced mechanisms in a later chapter, dry product testing. First we need to know how buoyant and saturated your App niche is.

Research your niche

Even if you only have one concept you wish to pursue, a little bit of time invested now could save you many thousands of dollars, pounds or camels later. If you have several concepts, then this will help you choose which concept holds the best opportunity for you commercially.

One of the most powerful ways to do this, is by measuring just how strong the demand is in your particular App niche segment. To achieve that I want to introduce you to an amazing tool that gives as hard statistical factual information on any market segment you can imagine, it's called the Google Keyword Tool. Internet marketeers live by it. There are other solutions, however this tool from Google is free to use for anybody with internet access, and it is more than sufficient for what we need.

Put simply the Google Keyword Tool reveals how many times search terms are entered and requested from Google, per month, per year, locally, or world-wide.

Whilst the answers it reveals are purely related to internet search requests, it seems appropriate that the level of searches carried out on the web may suggest similar behaviour and search requests on the App Store.

Get your book bonus material (includes Keywords exercise A4 work sheet): www.AppManSecrets.com/RichAppPoorApp.html

Keywords

List as many relevant keywords, associated with your as you can, aim to have at least 20

What follows is a series of easy-to-do exercises, that will give you valuable insights into your own App niche or niches. First we need you to list the associated keywords and short phrases related to your App niche are subject area.

Again this doesn't require lots of research, simply write down all those you can, we need them for the following action, you can use the box provided for convenience.

Using a web browser, search for the term *Google Keyword Tool*, and *click* on the appropriate link. It's likely to come top in the search, particularly if your using Google to search with, for obvious reasons.

Once you're into the website, simply type in the keywords, or short phrases you have identified directly into the panel provided. Be sure to put a *return* between each keyword or phrase, this tells Google it is a separate phrase.

Click on *Search* and hey presto, a list will appear with all the terms you have suggested, plus many others that Google feels is related to these terms.

Unless you have changed the setting, you will see information such as competition, Global Monthly Searches, Local Monthly Searches, Local Search Trends.

Google Keyword Tool

This tool has been created to help Google marketers advertise their products using Google's Pay-Per-Click (PPC) campaigns, which they call AdWords.

We don't actually need all this information, so to simplify the view go to the *Columns* button, which is on the far right of the screen, just above the Local Search Trends, *click* on the button, and turn off all the options so that just Global Monthly Searches and Local Search Trends are checked.

What you are left with is a list of the keywords, and phrases on the left, an indication of the Global Monthly Searches in the next, and the Local Search trends. Since you are not making an AdWords campaign, the individual keyword search term results are less important than how this niche performs overall, to assess that column needs to be added up.

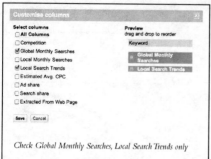

Check Global Monthly Searches, Local Search Trends only

Do note that there may well be more than one page of results. To add these up, you may either get your calculator out, or take a short-cut...

To save yourself time, and also for future reference which is worth having, you can export this data for future use, and reference. You can also add this data to a spreadsheet if you have one and get the software to add up that column for you. *Excel* will work perfectly on Windows, or if you prefer Apple's equivalent use *Numbers*.

To do so, simply *click* on *Download*, which you can find on the left above the keywords. You can select from the various options, such as *CSV*, and input into your preferred spreadsheet program. You can fine tune the terms you download by ticking the boxes, or you can edit out unsuitable terms within the spreadsheet program. Either way now you will have an accurate, unbiased, scientifically calculated number of people that are searching for those collective terms across your niche.

Since it is easier and better to assess business opportunities on a yearly rather than monthly bases, adjust this number to obtain the approximate size of your niche per annum.

You now have a very good indication to the size of your niche. There is no magic number here, in most respects the more the better, however this doesn't account for Competition.

Competition

There is a competition column within this tool, but that indicates the level of competition there is for those search terms on a PPC campaign, which isn't relevant. If you do look at Google's Competition data, then you actually want there to be the highest level of Competition possible, the greater the level of competition, it implies the greater the level for commercial opportunity.

What you want to be aware of is how much competition already exists within the App Store, which is easy to assess by going there and carrying out a search. If you don't already have itunes installed, you must do that first. Carry out a simple Google search to find and download this application from Apple.

Once you have iTunes installed, use the Keywords you have already established and carry out searches inside the App Store to discover how many Apps already perform the same task. Now divide the number of Global annual searches you identified earlier by the number of applications that are already performing the same task.

Is competition a good thing

There are two opposing view points, if you find that there is existing competition in the App Store that implies that there is a viable market in your niche. The bigger the size of the niche, the bigger the competition we can assume to find inside the App Store. So on one-hand competition is a good thing, and may even provide a degree of comfort, knowing others are in the market.

However if you have goals to dominate your niche, then any sign of competition might be seen negatively, and a potential barrier to you achieving your goal of niche dominance. No view point is right or wrong, and the results are open to interpretation.

Dig deeper

If there is competition, dig a little deeper. Are there any user reviews to learn from? Are users requesting features that aren't yet available? Are there insights to be learned?

Consider the following points: are these Apps making good money? Have they left an innovation opportunity that you can capitalize on? Can you offer something to this market that isn't currently available?

Competition

Are there any features not currently being delivered by the competition?

Are the graphics, services targeted to specific users, or just generalized?

Is there a sub-niche that could be targeted?

Is the competition providing solutions in all comercially viable languages?

Can you provide an improved graphic user interface, or improved user experience?

Is there a revenue stream that is not currently being offered, such as in-App purchasing?

Can you monetise your App using a different revenue system that may be have more commercial potential that those already existing. If the answer is yes to any of these, then you may have found the App niche for you.

Remember, many successful business owners don't like being first to market, they prefer to be second, and better. Learning from those that went before them; improving on the good work already done.

You can of course repeat the simple process of checking the commercial viability of your App niche with all App-ortunities you have already identified. If you do discover several Apps within the same niche, then this process can take a little time. However it is always time well spent, only the novice Rookie doesn't investigate their opportunity before leaping in. There are so many great opportunities in the App Store, there simply isn't any reason to leave it to complete luck. A little sound market research will pay off in the long run.

The App plan

In broad terms the approach to creating an application is easy, as you can see from the diagram below, dealing first with the concept or idea-ation, then the Apps development, and then releasing that application via Apple to the world.

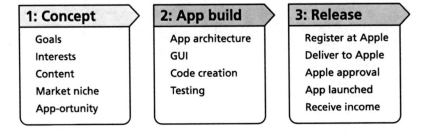

1: Concept	2: App build	3: Release
Goals	App architecture	Register at Apple
Interests	GUI	Deliver to Apple
Content	Code creation	Apple approval
Market niche	Testing	App launched
App-ortunity		Receive income

Concept: This is the initial step, coming up with opportunity in the first place. This book will help you identify that, with various exercises, looking at your goals, interests, content, market-niche, and ultimately discovering app-ortunity.

App build: Deals with the App architecture or navigation, Graphic User Interface (GUI), then the coding itself before thorough testing both internally, and at the clients end.

Release: Must register with Apple. This gives you the ability to control price, receive payments, adjust screenshots, App details and monitor sales. After delivering to Apple, Apple review each application in-turn before submitting their approval or rejection (which we discuss in a later section). On approval the App is immediately released, unless you have requested otherwise, the sales are monitored by Apple and your revenue is delivered automatically every month directly into your bank account.

The Rookie will take these requirements one at a time, however because most creators want their App in the App Store generating income as swiftly as possible, this is the wrong approach because some elements take longer to achieve than others.

Instead when we understand that different elements take different quantities of time, you can structure and plan the application much more efficiently.

In the graph below the horizontal distance of each section indicates how long that element takes. Whilst the concept phase is arguably the most exciting, apart from the launch that is, it generally takes the least amount of time.

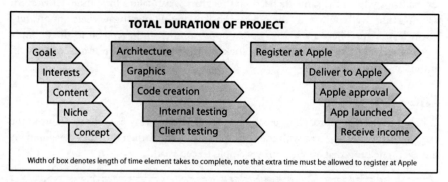

TOTAL DURATION OF PROJECT

Goals	Architecture	Register at Apple
Interests	Graphics	Deliver to Apple
Content	Code creation	Apple approval
Niche	Internal testing	App launched
Concept	Client testing	Receive income

Width of box denotes length of time element takes to complete, note that extra time must be allowed to register at Apple

With the help of this book, and the exercises within it, that element is even more efficient. The graphical and App architecture stage of the application is the longest, with the building phase being slightly shorter, but with one exception...

Registering at Apple can take a long time, it shouldn't, but it can. My strong advise is to register for the iPhone Developer Program as soon as you recognise that you have a solid concept, and that you are committed to launching your own application.

What is often under-appreciated is that you are about to launch an international product, and as you might imagine that involves some paperwork. Fortunately Apple have done a great job of streamlining this process for you. Basically they have already established the agreements with each Country, and you just need to form an official agreement and Register with Apple.

When enrolling in the iPhone Developer program, be sure to register as an individual, unless you have plans to become a commercial iPhone and iPad developer yourself

This is called an iPhone Developer Program, and you can enrol under either of two options, as an Individual, or as a Company. You only need to enrol in the later, if you intend to become an iPhone and iPad developer yourself, for all other situations simply select the Individual program.

There are several forms, completing one at a time before you progress on to the next one. This includes tax details etc. and these forms can take some time to get approved.

Doing this enables you to take control of your application, some developers offer to launch the application under their Company account, however with your own registration, you can take control, change prices, adjust sales text without having to go running to your developers each time. The annual cost is not prohibitive so there is little reason not to take control over your own products.

The registering process takes place digitally using a web portal from Apple called iTunes Connect, this may all sound very complicated, but it is actually very easy. Simply follow the instructions provided, ticking the green boxes as you move along, and you will be enrolled without any headaches. The headaches only occur, when you leave it until later in the process, read the *Lesson...*

Lesson...

One customer had worked with their application's development. Whilst the team were busy creating the code, and delivering visuals for approval, they had established several joint ventures, and advertising agreements, including a national mobile phone Company.

The application's development went as planned, and after the App was satisfactorily tested on the clients machine and approval received the application was 'built' ready for uploading to Apple.

It was only at that point that the developers discovered that the client, had not completed the required paperwork.

As a result nothing could be done, and the developers were powerless to upload the App until all the paperwork was compiled, delivered and approved by Apple. In the mean time, the advertising agreements, and joint venture arrangements were jeopardised, deadlines missed, and launch events cancelled.

So please learn the lesson, get registered at Apple with iPhone developer program, which does cover development of the iPad despite the name, and ensure that all the documentation is complete.

By now, not only have you identified your own App-ortunities, you have assessed the size of that potential in commercial terms, and have annalised the correct competition. By completing this process across the opportunities you have identified, you will greatly enhance the commercial reward you are likely to receive from your application once fully developed and released.

In the following chapters, you will build upon this concept, developing it further, and refining all aspects to ensure that you maximise your App-ortunity.

Know your customer

Apple's huge financial success, is because they are providing an environment where people can buy and sell highly specific micro-niche products. This trend is so strong that Apple are recording their biggest ever quarterly profits when the Banks, the Airlines and many other huge institutions are gong bankrupt or announcing huge losses.

To produce services for these micro-niches, we need to forget about thinking of your users as a huge diverse collection of customers and start thinking about them as unique individual people, with highly individual needs, desires and unique personalities. "Emotions enrich; a model of mind that leaves them out is impoverished" Goleman 1995. This may seem counter intuitive, but then the best insights often are. So let me show you why it works this way...

> **❝ Emotions enrich; a model of mind that leaves them out is impoverished ❞**
> Goleman 1995

Example
If you went about designing for a 40 year old man, the 20-something race car enthusiast and a mum all at the same time you might end up with some awful hybrid monstrosity.

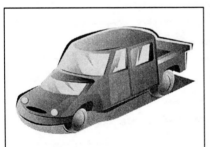

But if you *tapped* into a specific personality, context and individual, you would end up with a highly targeted product that would fit it's market perfectly, such as a sports car, or a 4-door saloon etc.

There are many ways to research and understand your user. I explored this in great depth in my first book *Make Your product Invisible - or nobody will buy it*. Here I explained not only the research techniques being used by the leading future-shapers of our time, but also techniques I had developed myself which are aimed directly at a society we are all moving towards.

This isn't a design course, or a course in entrepreneurship so I'm not able to cover that here, however what I will share with you is a very simple technique called Story-Boarding. It's simple, but leads to great insights about who the customer is, their situation, their context and their mentality. *It's fun too...*

First of all we need a star, every good story has a star, you get to choose your own...

My ideal user is...

Remember there is no way to do this wrong, it's your story so you can choose who is the star of the show. So describe below one unique ideal customer for your App...

Male or Female _____

Age _____

Nationality _____

Name _____

Are they working, at school _____

Describe their work, school enviroment

_____

How much do they earn per year _____

What are their main hobbies, their passions, indulgencies _____

Where do they shop _____

What car do they drive _____

What's their favourite band _____

Where do they live _____

Who do they live with _____

Know your customer: Exercise #2

Context

Consider where they might be when they first decide to search for your application...

Where are they

_____
_____

What are they thinking about, when they first search for your application

_____
_____

Describe their emotions

_____
_____

What do these emotions cause them to search for, what do they want to find, what are they looking for

_____
_____
_____
_____
_____

Sketch

I realize that not many people like to sketch, however this can also eccentuate the appreciation for the customers situation, so have fun, and give it a go even if your not a fabulous artist...

Sketch the situation they are in, when they decide to search for an App solution...

Sketch their situation when they use the App for the very first time...

Now sketch the situation, another time they are using the application, in a different environment, maybe even for a different reason...

Know your customer: Exercise #4

Describe...

Thinking completely and only about your user, answer the following...

What does the App need to do

How should it make them feel

How should the App look

What experience should the App deliver

What is the benefit to them in owning the App

How will they know when they have found the right App for them

----

Make sure you give this information to your developer, it gives them huge direction, and helps them greatly, and will enable them to do a better, and quicker job for you.

App Principals

There are five App revenue models...

Direct App sell

The most obvious method. With the help of a great Developer you create a Great application, which you sell on Apple's App Store for a flat fee!

All the payments from your customers are paid directly to Apple, who will then retain 30% commission for their worldwide distribution and deliver 100% of the rest to you into your bank Account every month!

In App purchasing

Apple introduced *In-App-Purchasing* for the first time in late 2009, and it provides a fantastic new Business Model to App creators.

As the name suggests, users can *tap* on a selection that will take payment from them, this is a great way of getting your Customer to buy more packs, levels, features, or content from within the application itself.

For example a game can use *In-App-Purchasing* to get you to buy and download the next five levels of a game. A magazine could do the same with extra articles etc.

You can request monthly, yearly or periodic payments for your users to acquire new content. You get paid in exactly the same normal way straight into your bank account on Auto-Pilot each month.

This is a brilliant way to keep your customers engaged with your App and drive additional revenue at the same time. It's always easier to sell an existing customer than finding new business, so it's a great technique. In my opinion this technique is still being under-used and provides a superb opportunity.

Subscriptions model

This is where a regular payment is set-up and paid for at regular intervals. This allows the user to continue using the App service, or provide them access to new content that may or may not be provided. For example the first Turn-by-turn Navigational iPhone App was sold at $9.99 per month.

It is important to realize that Apple does not at present allow indefinite monthly plans instead the user must commit to a fixed term payment plan, with the option to renew at the end of that period.

So for example you could choose to purchase a subscription based App for 12 months with regular monthly payments coming out of your Bank account via Apple's iTunes system. At the end of that 12 months you would be invited to purchase another subscription, if you choose not to then the App would become un-usable.

This is a great Business model, particularly for high end solutions that require higher price points to fund their development, whilst at the same time providing a bite sized price per month rather than a huge one time payment upfront, before you have even tested it out for real.

Free App

I'm often asked why are there so many Free Apps, and how do they make their money.

To understand this we need to take a look at the impact an App can make. Let's compare it to websites for example. Websites have become so valuable that the first thing almost every new business does is to check and buy a suitable domain name.

So let's assume you can make more impact with an iPhone App than a website, but how much more?

	5 times?		10 Times?		50 times?
	100?		1000 times?		10,000 times?

In July 2010 there are 200,000 Apps available in the App Store, there is however over 1 trillion website's according to Google's data in 2008. So every App makes seven million times more impact than a website does, and that is not a typo.

> ❝ how much more impact does an App make compared to a website ❞

When you have that kind of impact, and that kind of traffic, and eye-balls it's very easy to monetise it through advertising, or by up-selling to other products and services. So there are five types of App business models-

> **Sell** – via one time payments
>
> **In-App Purchasing** – to sell them additional features, or content
>
> **Subscription** – incremental payments over a fixed duration, and
>
> **Free** – using Adverts to provide income or up-selling to other products or services. The fifth option is...
>
> **Mix and match** – you can of course use several of these techniques within the same application.

None of these systems is better than the other – all can be viable, depending on the nature of your application.

Why Apps get rejected

Apple have produced a document for developers called *iPhone Human Interface Guide-lines* which is basically a list of Apple's design principals.

It's a huge document, but the good news is you don't need to read it. Instead what you do need to do is use a specialized Apple Developer like us who work with it every day, and know it inside out.

What's more important is that you learn why Apps get rejected so that yours doesn't!

No.1: Technical issue

If your App does get rejected, you can adjust and re-submit, each time it takes time, and possible money to make the App adjustment, and create the new build for submission. By using a specialized and experienced App developer you can avoid these issues, because they know Apple's requirements.

No.2: Adult material

With regards to adult material Apple have adjusted their policy, and at the time of writing only very light nudity or suggested images are deemed acceptable. If you have an App based around such material you will need to be very careful to get your App into the App Store.

No.3: Violent material

If you have an App that promotes violence, then you will need to tread care-fully. The best thing to do is examine Apps currently in the App Store and use that as your limit to the quantity of violence that will be tolerated.

No.4: Immoral

In the early days of the App Store there was an App called *I am Rich* which sold for $999 US Dollars, it contained just an image of a Red Jewel!

It didn't contravene any of Apple's protocols, however after many complaints it was withdrawn from the App Store and no longer available for purchase.

A good developer will provide advise here, and whilst Apple are free to adjust their own policies any time they choose, a professional developer will at least notify you if your material is approaching current Apple thresholds of descent.

Learn from Apple...

According to Steve Jobs the founder and CEO of Apple, Apple are primarily a software Company, they package that software in beautiful boxes, but at the end of the day they provide software solutions.

Apple are arguably the best design Company on the planet, with both their hardware and software, winning many International design awards.

Apple have won more prestigious D&AD black pencil awards than any other Company in the world

Rather than learn everything there is to know about Apple's Human Interface protocols, it's better to simply use and have a basic understanding of how Apple's own Apps work.

This provides evidence of what the best software writers in the world are producing; an insight into the Apple App logic; and most importantly the Apps they produce such as *Mail, Safari, Phone, Calender, Camera, iPod* and *Contacts* are the Apps your customers will be most familiar with.

So it makes sense to use the same principals your customers are already familiar with, which is why we are going to take a look at their key applications...

Calendar

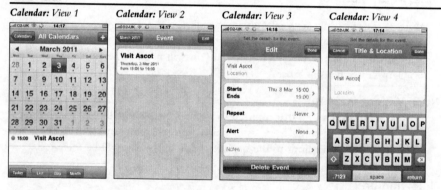

Calendar: View 1   *Calendar: View 2*   *Calendar: View 3*   *Calendar: View 4*

View 1: Shows the main over-view page. Whilst there are many areas to *tap*, and therefore many actions the, page looks simple and easy to understand. Tapping on any date reveals the logged events of the day which are listed under the calender. Tap on any item under the calender to examine it further.

View 2: This shows a summary of the event selected from the previous screen, note that no information can be edited on this screen. The only options the user has is to either return to the previous screen by *tapping* on the arrow top left, or *tapping* on Edit to make changes.

View 3: Here the user can select any element to make changes, or to delete the entire event by *tapping* on the red button.

View 4: Tapping on any element on the previous screen opens the keyboard view to easily make changes.

Note that it is possible to design a calender application which can be viewed and edited in fewer steps.

Photos

Photos: View 1 *Photos: View 2* *Photos: View 3*

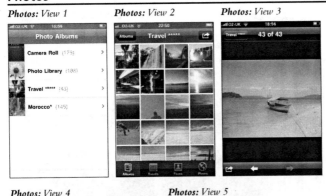

Photos: View 4 *Photos: View 5*

View 1: Shows the available photo albums.

View 2: Shows the images contained within the selected photo album, in a neat square icon view. Note how no image names, file size, image type, date or associated information is revealed here.

View 3: Select any image and it will be displayed in full view.

View 4: Rotating the phone will resize the image to fill the screen, note how there is no button, icon, explanation or help page to indicate this action exists. It is worth pointing out that if you rotate the phone in other applications such as *iPod*, very different actions are performed.

View 5: Selecting the bottom left icon shown in *view 3* will reveal image export options as shown in *view 5*.

iPod

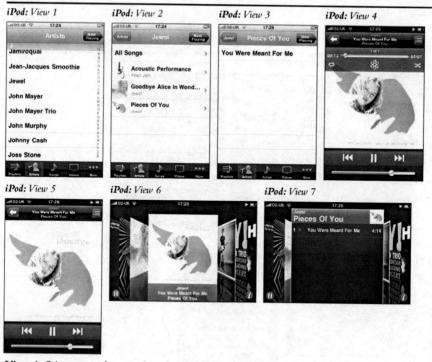

iPod: *View 1*

iPod: *View 2*

iPod: *View 3*

iPod: *View 4*

iPod: *View 5*

iPod: *View 6*

iPod: *View 7*

View 1: 5 icons on the panel at the bottom of the screen provide swift and easy access to key areas of the application. The example shows the Artist view, with the Artists clearly listed in alphabetical order to aid selection.

View 2: Tapping on any artist will reveal the albums related to that artist.

View 3: Tapping on the album of your choice will reveal all the songs within that album, in this example there is just one related song.

View 4: Selecting any song in this way will start playing the chosen song, revealing the graphic, and various buttons and options to manipulate that selection. Options include launching *Genius* moving to a different area in the track, adjusting volume, selecting the shuffle song option, repeat options or skipping forward to other tracks.

View 5: Tapping the screen again will hide most controls to make a less cluttered screen. Note how there are no indications that this action exists.

View 6: Rotating the phone, will reveal a completely different way to navigate and select different songs. Note how again, there is no indication to reveal that this significant feature exists.

View 7: Tapping the album cover reveals tracks within that album, again there is nothing to indicate that this feature exists

What can we learn from Apple?

Clarity not speed

Can you see how Apple has made every step very clear. It leads the viewer through several screens sometimes, but always provides clarity, and the user always knows where they are. This is particularly in evidence in the *Calender* application, where fewer steps could have been created if Apple had wished to do so.

So why has Apple done it this way, did they make an error? I think not!

Not a website

I see this Rookie mistake all the time. Rookies assume the Apps are mini websites, but in-fact they are their own specific animal. Websites usually search for the fewest *clicks* to navigate to any location possible. Yet we can see by Apple's own App examples that this is not the preferred approach.

Instead they would rather sacrifice a few extra *taps*, to keep the navigational structure clear and uncluttered wherever possible.

Web experiences are generally performed using the mouse which is a brilliant device, thats become almost second nature to us. However, compared to a finger, it will always feel primitive and unintuitive.

As the Authors point out in the book *iPhone User Interface Design Projects* the dexterity you have in your finger is so quick, versatile and accurate that its incredibly easy to make multiple *taps*.

If you attempt to reduce the *click* required in your App then the interface can easily become complicated, and the user can start selecting elements by mistake and get frustrated with the over-all experience. As those Authors put it *'Taps are Cheap!'* so do not sacrifice clarity in the search for fewer *taps*.

Clean

We can see from Apple's Apps, they use images a great deal, and seem to refrain from displaying words wherever possible. This certainly leads to a more beautiful visual solution.

Apple does have one big advantage though, users will invariably be using Apple's core Apps on a very regular basis, and as such, even if we are not familiar with a few icons initially, it wont take long for us to learn.

So whilst generally we all love having beautiful uncluttered, clean screens without text, it is important that the user is clear what the various buttons and options available to them do. Sometimes the best, and fastest way to achieve that is by using words. So I suggest that we need to use words and or images that describe the function best to the user, even at the expense of a little beauty.

App case studies

Applications are categorized into various classifications for easier referencing. In the previous chapter we took a look at great examples of Apps in the Productivity, Photography, and Music categories. In this chapter we will continue the investigation, this time using non-Apple examples across seven other categories: *Games, Utilities, Lifestyle, Reference, Navigation, Health & Fitness,* but we are going to start with *Finance.*

Finance

For the record neither *Credit Cool* or *Account Tracker* have any relationship the Author in any way *[Editor].*

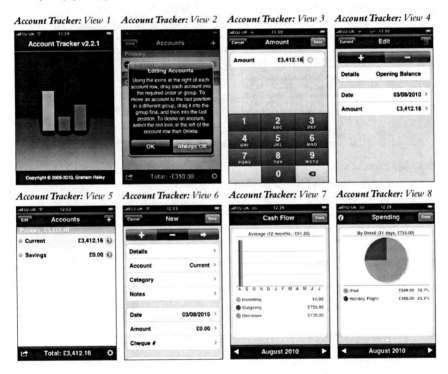

Account Tracker: View 1   *Account Tracker: View 2*   *Account Tracker: View 3*   *Account Tracker: View 4*

Account Tracker: View 5   *Account Tracker: View 6*   *Account Tracker: View 7*   *Account Tracker: View 8*

Account Tracker allows the user to track their various bank accounts, logging date and category details every time the user inputs their financial transactions. This data can then be displayed through a number of views to track the users expenditure, and feedback purchasing patterns.

This application utilises familiar buttons, colour schemes, and page designs, within a bespoke App architectural structure.

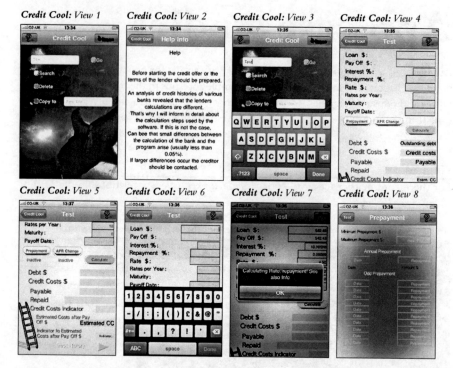

Credit Cool: View 1   *Credit Cool: View 2*   *Credit Cool: View 3*   *Credit Cool: View 4*

Credit Cool: View 5   *Credit Cool: View 6*   *Credit Cool: View 7*   *Credit Cool: View 8*

Credit Cool allows the user to input and track the payments of any loans, allowing the user to have a permanent digital record of payments at all times for convenient reference. This application uses bespoke screen design, buttons, and icons within a non-standard App architecture.

Even if you have never used either application yourself, do you have a preference?

Which do you think will generate the higher income?

In tests using these two examples, I have yet to find anybody that preferred the *Credit Cool* to the *Account Tracker* application. This is reflected in the applications commercial performance, with *Account Tracker* being towards the top of it's App category, and *Credit Cool* floundering around the bottom without any significant sales.

The difficulty with *Credit Cool* is not with the function it performs, the ability to track loans will for some be a very beneficial aid, it's the execution. The overall user-experience is so poor, that very few users are going to be pleased with their download, learn how to use it, recommend it to others, or give it a positive review on the App Store.

This is an extreme example I admit, however it proves the mistakes that Rookies are making when they do not get a proper developer on-board, or fail to obtain any expert advise. Place the same application concept in the hands of a competent developer, and the results will be massively improved.

Here is a list of Credit Cool's Rookie mistakes...

✗ No expert advice

✗ No concept testing

✗ No user research

✗ Didn't use a professional developer

✗ Rushed in without advise

✗ No feedback loops

✗ No way to capture users data

✗ No strategy for positive user reviews

✗ No product funnel

✗ No focus on the objective

Credit Cool recommendations...

✓ Start again

✓ Use a great developer

✓ Understand your audience

✓ Produce a story board

✓ Produce great Graphical User Experience (GUI)

✓ Approve 2D version, before developing App build

Whilst Account Tracker, is clearly a far superior application, there are a number of ways this could also be improved...

Account tracker recommendations...

✓ Include feedback loops

✓ Introduce product funnel

✓ Develop pro-version

✓ Video support

✓ Compelling offer which can be used to capture valuable user data

✓ Word-of-mouth. Use mechanism to encourage people to talk about application to others

✓ Define the ideal user, and shape the application more specifically to this user's needs

NightStand – Utility

NightStand: View 1

NightStand: View 2

NightStand: View 3

NightStand is the only application listed here that has been developed solely for the iPad. This is a very simple, but elegant application which simply displays the current time in a variety of styles in full-screen. The user can adjust the screen brightness, and as the name suggests can be used as a night light for safety, or for comfort. In addition it contains the usual alarm options you might expect, including the option to awake to morning bird-song, which is a nice addition.

TapTap – Games

TapTap: View 1 *TapTap: View 2* *TapTap: View 3* *TapTap: View 4*

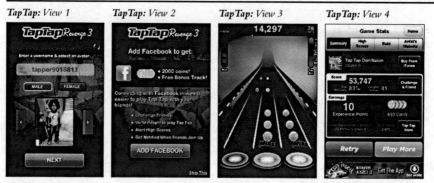

TapTap is a fast paced music *tapping* game. The user must *tap* on the areas at the bottom of the screen in-time to the music, as indicated by balls and various objects and images that *fly* down from the top of the screen.

The user is encouraged to connect to Facebook, and is rewarded with various bonuses if they do, which also enables the application to gain further exposure.

Trism – Games

Trism: View 1 *Trism: View 2* *Trism: View 3* *Trism: View 4*

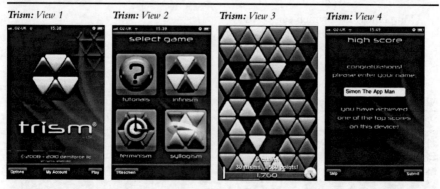

Trism is another game. This one is time based on grouping objects of the same colour together which then deletes them, and replacements slide into place depending on the orientation you are holding the iPhone in at that moment.

Lottery – Reference

Lotto: View 1   *Lotto: View 2*   *Lotto: View 3*   *Lotto: View 4*

Lotto is a reference application that displays recent National Lottery results in the UK and EuroMillions results from Europe. Users simply select either Lotto or EuroMillions panel from the bottom of the screen, and *swipe* left and right to select the draw they want to see the results of.

iFitness – Health & Fitness

iFitness: View 1   *iFitness: View 2*   *iFitness: View 3*   *iFitness: View 4*

iFitness: View 5   *iFitness: View 6*   *iFitness: View 7*   *iFitness: View 8*

iFitness is a feature rich fitness application. Users can select from a wide range of exercises that are categorized into muscle groups. They can then see images of the exercise, and videos that demonstrate how to perform the manoeuvre correctly. A number of additional features are included such as fitness schedule, weights, timer, Body Mass Index etc.

So which App do you prefer?

☐ NightStand?   ☐ TapTap?

☐ Trism?   ☐ Lotto?

☐ iFitness?

When we have asked the same question at live events, *TapTap* or *Trism* gets the winning vote. We investigated why, and asked participants to evaluate all these applications on App name, Logo, Clarity of task, Introduction/Greeting, Graphics, Price-Point, Navigation, Help support, pat-of-product-funnel, Build a database, Word-of-mouth, and Feedback loops, we got the same consistent results.

What we discovered

It is worth noting that some applications are heavily feature rich such as *iFitness*, whilst others are minimal such as *Lotto* and *NightStand* so this is the not the key factor in achieving a leading mobile application.

So it is not necessary to create the most complicated feature-rich application you can possibly put together. It is much more essential that you concentrate on usability and a clear navigational structure.

The number one element that interested our audience was Graphics, winning by a landslide. All the applications listed here are towards the top of their respective categories, and it is worth noting that all of these examples have done a good job of using clear and attractive graphics throughout their application.

If you compare the graphical execution of these compared to *Credit Cool* mentioned earlier it is very clear that their investment in good graphics has been a wise one.

Successful

All these Apps have been highly successful. *Trism* generated over $250,000 in two months, *NightStand* for example made over $20,000 in just it's first three weeks of the iPad being on sale (and was only on sale in America at that time), and *TapTap* had over 1,000,000 downloads in just it's first two weeks.

So again, you do not need to create a huge complicated application to obtain App riches, as these case-studies demonstrate.

Research your niche

If you have identified an App niche, and have some competition you want to assess, feel free to do so using this evaluation sheet.

Most applications are easily assessed once you have a clear structure to do so...

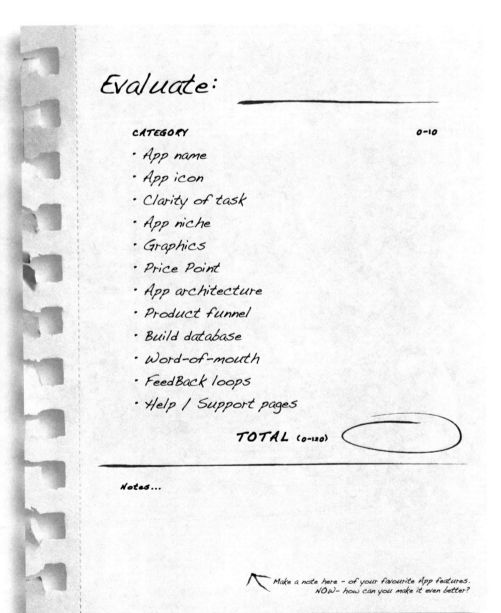

Evaluate: _____

CATEGORY   0-10

- App name
- App icon
- Clarity of task
- App niche
- Graphics
- Price Point
- App architecture
- Product funnel
- Build database
- Word-of-mouth
- FeedBack loops
- Help / Support pages

TOTAL (0-120)   ⬭

Notes...

↖ Make a note here - of your favourite App features.
NOW- how can you make it even better?

#RICHAPP

App architecture

With so many opportunities to control an application, such as gesture based controls, voice control, as well as the more familiar *taps*, there are infinite ways to deliver your application. However there are some key approaches to App architecture, or App navigation.

Most applications start with what is called a *Splash page*, or *Splash screen*. This is usually a graphic screen, with no functionality at all, that remains on the screen for just a few moments. This achieves a number of purposes. It serves as a nice graphical intro into the application, it is after-all your shop window to your own little App Store, and arguably more importantly, it allows the application to gather any data it needs to function in the background whilst this page is on display. This may take a little longer if the App has to access online services to draw down data, however in most cases this screen disappears within a second or two to reveal the heart of the application.

Navigation

Whilst there are thousands of applications, and therfore thousands of App architecture solutions, a huge majority fall into one of four distinct categories, which we refer to as *Swipe*, *Panel*, *Tree*, and *Custom*.

After selecting your App's revenue strategy, this is probably the aspect that shapes your applications form more than any other aspect. Getting this right is essential.

Of course, you can leave this to your development team, however in many cases they will search for the easiest solution to build, and that may not be the solution that your customers require, which is why it is important for you to understand the basic architecture principals.

Swipe: In our previous examples we looked at the *Lotto App* which is a classic example of the *Swipe* navigational system. To navigate from one screen to the next, the user literally swipes their finger across the screen. Small dots towards the bottom of the screen indicate how many pages there are in total, and which of those they are currently viewing.

Swipe based applications have been said to resemble digital cards. That is the user can swipe to change the card, but usually their is also a small button that will flip the card over to reveal further information.

This architecture is very simple, and very elegant. It is most suited to small applications that are displaying text or static visual information, with a limited number of pages in total.

Tree model: The name *Tree* is used to denote that the structure is like a family tree, or you could think of the structure as being like a tree's roots. As the user moves down the root, the user gets to more and more specific information.

This approach works well for news based applications for example, where the user may first select Sport, then Football, then Premier League, then Arsenal, and so on. At each stage getting to more and specific data or information.

This architecture is often used in larger applications, where lots of information is contained and the easiest way to provide navigation to all the elements is through this hierarchy *Tree* structure.

On the negative side the user may be required to *tap* many tines to get to the information they require, and may have to travel all the way up the root before they can travel back down another fork in the path, or root of the tree.

Panel: As your App graphics, and layout evolve you will realise how critical it is to have a clear navigational structure to your App. Particularly on the iPhone where space is at such a premium, having a tool bar of some description works well on the web where it is usually displayed at the top. On an iPhone where the hand is used to *tap*, the tool bar is better displayed at the bottom, so that the hand doesn't obscure the screen when selecting. Tthis tool bar is usually called the *panel* on Apps.

This allows the user to jump to significantly different sections with just one *tap*, whilst at the same time informing them where they are located at the time. Very simple, but very powerful. Even simple Apps with minimal content are often improved by using this standard mechanism. It's not very revolutionary, but it works.

Custom: There are of course many thousands of applications that do not possess any of these techniques, instead employing their own unique bespoke graphic user interface approach. In the best examples such as *Tap Tap*, this uniqueness adds intrigue and enjoyment to the overall application. However, by introducing new navigational techniques, you risk confusing your customer.

Generally speaking the most innovative architectural systems are used in games, where the level of interaction is typically at it's highest. Whatever the genre of App, a great deal of user testing is required with all custom architectural approaches. What may seem completely natural and intuitive to you, may confuse and annoy many potential customers.

You can assist the transition, or the education with help and support pages, using video wherever possible, because that is always the best way to demonstrate how to use software of any description.

If you are going to introduce unique controls, adopt the philosophy of *once learned never forgotten*. If your users will remember the actions they have to perform to carry out a action, then you are a long way there. Gesture based controls achieve this well, as long as you do not bombard them with too many inside the same application. Motion, and interactions seem to get learned much more intuitively than colours, icons, and language ever will.

Wherever possible you need to remove *thinking* from the users experience. They should be able to use your App, half a sleep, with one eye-open, whilst on the train looking at the pretty girl (or guy) opposite them.

If the user has to *remember* which icon means what, then it's not working well and you might do better just to use words like *Save, Help*, rather than sexy icons nobody can actually comprehend.

The rule here is not to make any assumptions and to test absolutely every aspect, to ensure you are going to deliver an application that works, is exciting and easy to use.

The last option: Is a collection of elements ands part of all of the above.

Don't sacrifice navigation for fewer taps

Many Rookies look for the optimum navigation to allow users to access the array of features available. In that effort they often become *tap* efficiency kings, often at the expense of navigation intuition.

Although stating the obvious, it is worth recognizing that iPhone and iPad users are using the best pointing device ever created, your finger. And whilst with a mouse there are a number of actions to manoeuvre a curser/arrow around the screen, the use of a finger could not be more intuitive.

Generally speaking, users are happy to *tap* a couple of times as long as it's clear where. So look to reduce screen clutter and make the navigation easy, linear, and simplistic, even if that requires an extra couple of *taps*.

Feedback loops

I have a big thing about feedback loops. I think they are vital. As I am sure you know, any user can put a review of your App on the App Store, and if positive will help greatly to generate further sales. But I ask you, how many times have you given feedback?

I have downloaded many applications, and I have yet to give feedback positive or negative on any App I've used. So it's clear: if you want to know more about what your users think, then you need to create better ways for them to communicate with you.

There are a dozen ways to achieve this, even a simple feedback button linked to the Mail App is a vast improvement. To go one step better, give the user a benefit for giving feedback. Corporations spend millions getting valuable user-feedback, you have the chance to do for peanuts, yet many Rookies don't take advantage of this. You could offer them a prize draw, a discount of your next App, find a compelling reason that a user will want to give you that feedback, and it will help shape the development of your App in the future, enhancing your product and your income in the process.

Dry product testing

In publishing, a *wet proof*, means a printed example that has come off the print line, the term 'wet' is used because the ink hasn't even had a chance to dry properly at this stage.

This is the ideal proof. It is an exact example off the press. Typically a Designer or Project Manager would approve the first good copy, before they continue to print the remaining copies. However the problem with wet proofs are that any errors are incredibly expensive to correct (trust me I know). As a result a work-around is usually found, or the error is accepted. Whilst this can be frustrating and un-ideal for a magazine run, it can be deadly for a business or any commercial product.

What you need is a way to test, as accurately as possible, without the time and expense involved in producing it for real. If this can be achieved then that allows the product to be evolved, enhanced, or adjusted to better suit the market which will lead to increased sales. How do we do that...

The answer is actually alarmingly simple... We just ask!

However we need to ask in the right way. A favourite quote of mine, from Henry Ford... "If I'd asked my customers what they wanted, they'd have said a faster horse." And he's right. So we need to ask better questions.

#RICHAPP

> ❝ If I'd asked my customers what they wanted,
> they'd have said a faster horse ❞
> Henry Ford

Rather than asking people what they want, we need to use better techniques to illicit valuable information from them. Ideo, the world's most renowned design consultancy, (Apple isn't a Consultancy) have amazingly revealed many of their research secrets, and have produced an amazing set of Method cards, each designed to get to people's underlying issues, thoughts, and behaviour patterns. You may be pleased to know there is an App for that, see the information panel below.

Although I would encourage you to look at these methods, or read my earlier book which discusses these techniques and others, they do not discuss Dry product testing, so that is what we will concentrate on here.

Before production

We do not want to actually create the full application before we can find out if people would buy it, instead what we can do is create an illusion of a completed application, and invite people to purchase it. With applications this is relatively cheap, and easy to achieve.

By looking at the App Store we can see that customers base their purchase decisions on screenshots and a text description, both of which are easy to create and illustrate.

You can easily create a web page, or blog page, put a good description of the App, and place some realistic screen-shots together. On the same page, display the anticipated price, with a clear *Add to cart*, or *Buy now* button.

Place a device on your internet page which counts how many people visit that web page, and how many people choose to *click* on the button (usually done by placing a similar device on the page that it connect to). You will then be able to determine what percentage of potential customers choose to buy the product.

> ❝ Creates a list of highly interested prospects, whom you can contact as soon as produced ❞

After the viewer has *clicked* on the *Buy Now* button, a page is revealed that explains it is a Marketing Research experiment which is looking to find out the interest that potential customers may have in your product. To thank them for their time and for expressing an interest in your application offer them the application for just 50% of the normal price once launched. All they need to do is give your their email address and you will announce the launch and how they can purchase at the highly reduced rate.

You can also request that give you further details, such as what elements they would like to see the App perform, what price point they are looking for, etc. The more questions you ask, the fewer responses you will obtain. So keep them limited to eight or less short questions, and make it as easy as possible for them to complete that questionnaire or survey.

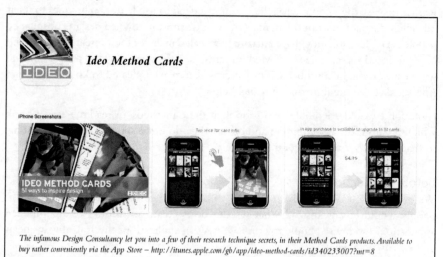

The infamous Design Consultancy let you into a few of their research technique secrets, in their Method Cards products. Available to buy rather conveniently via the App Store – http://itunes.apple.com/gb/app/ideo-method-cards/id340233007?mt=8

You do have the option of actually taking payment from them, the benefit of this is that you prove beyond any shadow of a doubt that people are happy to purchase your products online. However since most purchasers know that you can only buy Apps through the official App Store, they may feel uncomfortable actually depositing credit card details. In which case your statistics will become skewed, and not be a fair representation of demand.

The technique for actually taking money, even if only to return that to them is a valid one in some circumstances, although a suitable reward must be offered for their inconvenience, and disappointment. However it is better suited to non-App products, and therefore I would suggest it is unnecessary for your purposes.

Not only is this more honest, it also captures their details, and you know that you have an almost guaranteed audience on release day, which is a great approach.

This will achieve several key tasks;

1   Will tell you what percentage of people that viewed that page chose to purchase it

2   Creates a list of highly interested prospects, whom you can contact as soon as produced, so that you have an near guaranteed source of purchasers on day one.

3   Creates anticipation. When somebody decides they want something, and then can't have it, they will have a strong desire to close that loop, which they can only do by making that purchase. So they are likely to become hungry customers, and that's great news for you.

4   Has an opportunity for further feedback via on online survey for example, which may explain what they liked about your proposed application, what other features, benefits they would like to see. Preferred price point, do they mind adverts inside the App, what language they prefer the App to be in, etc.

Many companies would spend many thousands of dollars on such insights into their customers. By examining the answers, you can start to improve the application before it has even been created. Your App will have evolved on the drawing table, before it has ever been released, saving you valuable time, valuable resources, and your precious cash. *Genius!*

Traffic

Now all you need to do is drive traffic to your product testing web page. Luckily this can be easily done.

Social Media

We explore this in some depth later in the book, however it is clear that you can take advantage of any social online media to drive traffic to your web page very effectively. You can create a Facebook status update announcing your new App page. Your friends, colleagues, family will all be impressed, and invariably will take a look. Create a Facebook event, Facebook page, Facebook fan page, and you can take similar action with many of the other Social Media channels available to you.

Go one better

To go a step further, and to target your exact potential user you can create a PPC campaign, such as Google's Adwords, or a Facebook advert.

The benefits of a Facebook advert is it's easy to identify your target market through a person's indicated hobbies and interests. Once setup correctly you will only be charged when somebody chooses to *click* on your advert – which will mean they are a highly qualified prospect, and therefore worth paying for.

> ❝Just a few years ago these insights into your audience would have required a dedicated market research team, costing many thousands to complete. You can learn everything you need to know for free using online tools like this❞

PPC campaigns do incur some cost, so invest according to your App aspirations. Remember that big Companies pay fortunes to get this level of information. You are getting the same, if not better information for peanuts. However, balance that with the fact that all you need is some valuable customer direction, you don't need thousands of people to obtain that, so there is no need to invest heavily here and make sure you cap your account so that you will not spend a fortune here unnecessarily.

This book does not go into any detail of putting together an effective PPC campaign. There are many other books that could better advise on that subject.

Know your user

As a result you will get to learn your customer better. Ask the right questions, and you can learn the sex, age, income level, nationality, educational background, hobbies, motivation, passions, desires, challenges and keywords. The benefit is you will be able to enhance your product and ensure it fits your customers needs and desires *before* it has ever been released.

Creating a survey

There are many ways to create a digital survey, you could simply create a page in *Word* and distribute that via email, but a better method (which has the added bonus of not using Microsoft) are online survey Companies such as SurveyMonkey.com, and WuFoo forms.

SurveyMonkey seems to have gained the larger following, my preference is however for WuFoo forms, with its graphical user interface, I find it easier to use. However,

WuFoo: great online survey form which is easy to send via email, add to a blog post, a social media status update, or to position on your existing website

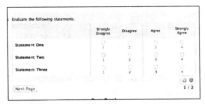

Click and drag, to add features, such as address, phone number, or even advanced elements such as likert scales

WuFoo, has a simple to use drag and drop format to create great forms in seconds. They will even send you an email every time your form is completed

You can use three separate forms, without any charge, all of which are easy to manage

Theme editors enable you to easily customize, and add background patterns for example. Some are better than others

either is completely sufficient for your purpose, and can be constructed in just a few minutes, *honest.*

Using a simple drag and drop GUI you can add address panels, email request, multiple choice, drop down menus, even advanced features such as Likert scales are all available out-of-the-box. You can have three of these forms fully working without committing to any contract, or paying any money at all.

The result is a quick-to-create solution, that can be created to your own specific needs, with a professional and slick look and feel, ready to be distributed via email, placed on a blog site, or embedded into your existing web page. *Easy.*

Just a few years ago these insights into your audience would have required a dedicated market research team, costing many thousands to complete. You can learn everything you need to know, for free using online tools like this.

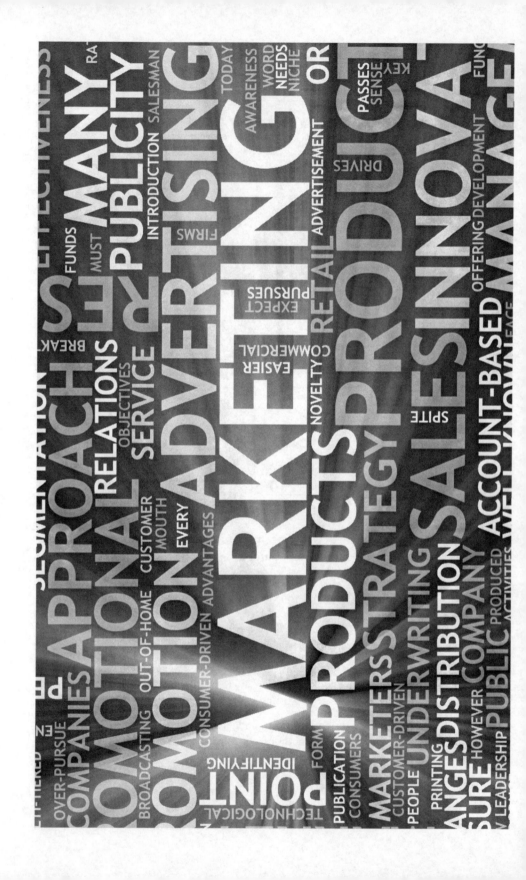

Marketing

The 2 sides of Marketing... / Marketing confusion

A great deal of mystery, confusion, and even suspicion seems to surround the world of Marketing. Those not familiar with, or without experience seem to assume that Marketing is some form of black-art, as if the skills associated with this valuable area of any business are beyond them.

> **66** Two critical elements to Marketing; Understanding your customers, and Asking for money **99**

Marketing seems to mean different things to different people. With so many unique interpretations, no wonder this leads to confusion. According to the The Oxford Modern English Dictionary, *Marketing*, *Marketing research*, and *Marketable* is defined as follows;

Marketing 1 tr. sell. 2 tr. offer for sale. 3 intr. buy or sell goods in a market. Be in the market for, wish to buy...

Market research the study of consumers' needs and preferences.

Marketable adj. able or fit to be sold

Whilst it is clear that Marketing is to do with the commercial opportunity, these definitions do little to help us. The exact boundaries of these terms are blurred at best. For example: does Marketing include areas such as pricing strategy, advertising, or adoption theory?

For the purposes of this book there are two critical elements to *Marketing* which we will define as;

Understanding your customers: Gaining insights into customer behaviour, user-interaction, customer requirements, opportunity gaps, commercial innovation opportunities etc.

Asking for money: Asking for the financial fulfillment of an order in whatever form that takes, including web transactions and purchases through the App Store etc.

Clearly understanding your customer is of critical importance. However this element is often over-looked in any business. This is a classic Rookie mistake that professional business owners irrelevant of industry have learned, often the hard way.

When you understand who your customer is, then it enables you to produce products and services to meet their requirements so much more easily, and so much more effectively. If you know what their habits and passions are then you can find ways to target them through advertising for example, or provide additional services that assist them in other areas of their life. To be commercially successful, there must also be a stage where the Company or the person supplying the product actually requests the order and asks for payment, or asks for the sale.

There are many forms of advertising, however most advertising is basically saying this is good, you should go and buy it. Adverts are basically asking for money, however dressed up the campaign may be.

It's clear that understanding either of these areas is going to be hugely advantageous for anybody selling a commercial product, service or experience. However to maximise any commercial potential, it is essential that both of these areas are done well. When both of these elements are aligned to create a clear coherent commercial strategy, then that's when the real magic comes.

As Chris West points out in his excellent book *Marketing on a Beermat*, there are a number of Marketing Myths, so lets bust a few of these myths before we progress deeper into this chapter...

Marketing Myth 1:

Marketing is only done by Big Business!

On the surface of it, this appears to be true. When I turn on my TV, or listen to my Radio I only hear about national and international products all created by big businesses. Well that's no surprise, no point a local decorator advertising on National TV when he can only service his local community.

However open the local trade Directory, or do a local Google search and you will find hundreds, or thousands of local Businesses or Tradesmen selling their services. Many of which would not have a viable business without the opportunity that provides them to attract new customers.

There is also a chicken and egg situation. Is it really more true that only Big Business do Marketing, or that you can't become a Big Business without Marketing?

All Companies and Products can grow in popularity via good word-of-mouth promotions. However, the pace of that growth is gradual at best and can be significantly enhanced if you can identify a way in which to introduce your products and Services to more people than those you have already worked for.

So if you have ambitions to grow the popularity of your App quickly then it stands to reason that some form, and appreciation, of Marketing is a good thing to acquire. Indeed, one of the biggest attractions to creating an App, is that a powerful worldwide Marketing platform is provided in the form of Apple's App Store.

Once upon a time everything we purchased was produced by huge international conglomerates. However, devices like the internet and the App Store are levelling the playing field allowing individuals to have access to the same exciting Markets that have previously been reserved for the multi-nationals. Consumers are searching for more bespoke solutions, which is taking us away from the huge multi-nationals, although do realise that you will soon be a multi-national yourself!

> ❝ The reality is good marketing works,
> bad marketing doesn't. ❞

Unilever may well have a £10 million marketing budget to launch or re-launch a product, but small business, and Entrepreneurs like you have more marketing potential then anybody in your position has ever had before. This really is a magical time for individuals in business.

Marketing myth 2:

Marketing is a con

Marketing is simply the process of connecting a person that is looking for a Product, Service or Experience, with those that deliver it. Any sense of 'Con' comes from a disparity between what is advertised, and what is delivered.

Whilst I cannot deny that there have indeed been many incidences of fraud in many industries, I cannot find any case of an App committing fraud within the App Store.

The closest example I can find is an App called 'I'm Rich', which was on sale for $999. The whole point of this App was that owners could show off their wealth by owning such an expensive application, which simply showed an image of a jewel, and nothing more. In fairness to the creators, they never actually mis-represented the application in any way. However, not surprisingly they received a multitude of complaints, and it wasn't long before Apple took down the application.

Sales or marketing lies usually occur when a sales-person who doesn't have the control to change the product is attempting to sell to someone looking for a different solution. If you discover that your audience are looking for different features, then you have the power to change the product and include these elements, usually without major expense, so their simple is no reason or benefit to lying.

Marketing myth 3:

Marketing doesn't work

If that were true Unilever wouldn't spend £10 million on an advertising campaign to re-launch Pot noodles. Nike, Adidas and all the other house-hold names would save themselves millions.

If that were true, Nike wouldn't invest 1.74 billion in advertising, (that's not a misprint), investing $20 million on Tiger Woods alone. They invest so heavily because it works, it produces results and increases their profits. The reality is good marketing works, bad marketing doesn't.

9 reasons for an App Marketing strategy

1 Knowing your customer
Knowing your customer is critical for any successful product, which is why we addressed this in detail in our earlier chapter. You may not have realised it as you carried out the exercises, but you were in fact carrying out sound marketing studies.

2 Allocate Marketing budget
By having a targeted Marketing strategy you can avoid Marketing channels which are not in-line with your goals, and outside your ideal user's areas of interest. The more scattered your marketing campaign is, which happens when you don't have a clear and focussed strategy, the less effective your time and money will be, and the worse the results will be. A clear targeted approach will maximise the return you get for your time and money (if any) invested.

3 Measure your progress
You will not be able to assess the effectiveness of your marketing unless you have a way to measure it. Some channels are easier to measure than others. If you place a banner over the motorway announcing your latest App it will be very hard to track any growth in sales, apart from comparing pre-banner sales, to post-banner sales.

However, this is never a great way to measure because other factors can come into play to distort the figures, such as time of year, whether, another App launch etc. Wherever possible track directly.

You can track how many emails are opened, you can track how many people *click* on a web page, or how many people sign-up to a web squeeze page which is designed to capture data. You can track how many people are *clicking* on a web banner, or an in-App banner from another application, then use that data to influence your future marketing behaviour.

4 Obtain funding, establish Joint Ventures
If you wish to obtain funding for your App, then any potential investor is going to want to hear about your marketing thoughts. If you can demonstrate a focussed and targeted Marketing campaign, that will make the process a lot more appealing to any investor.

It's the same story if you wish to do a joint venture with someone. The more information you can give, the easier it is for them to invest.

5 Marketing timing

With some planning, and a little forethought, it is possible to plan a launch campaign. It is worth pointing out that you can submit your application to Apple in advance to ensure it is approved, and then request that delay inclusion into the App Store until a specific date.

Whilst it is tempting to get the App launched as soon as is possible, with some coordination, much greater excitement can be generated. App reviews, press releases, social media updates, blogs, magazine articles, emails, twitter announcements, advertising campaigns can all be aligned to create real excitement and interest on day one of the App being available. This would not be possible without some commercial strategy, or marketing plan in place.

6 Sales targets

Sales predictions are notoriously difficult to achieve, however with targeted marketing comes improvement. Whilst our assessment of your market niche using Google Keyword Tool we explored earlier in the book gives an indication to the size of your market, it cannot predict the exact amount of App sales you will achieve in any one day.

However, if you were to launch an email campaign for example going out to 10,000 people, it may be easier to estimate the realistic conversion from that email campaign into sales, particularly if you have done similar campaigns before.

7 Evaluate your competitors

Surprisingly this is often neglected, which is why this book gives an exercise in *Know your niche* to encourage you to do exactly that.

We explored the App Store for your rival competition in an earlier chapter which gives you a great idea of your current competition, if any. If you want to stay up-to-date with App developments then take a look at *www.AppTism.com*, you can create an App watch list there, and it will inform you of any App updates for that list of Apps.

This way you will always be informed of new developments, announcements, reviews in your App industry. You could for example monitor a particular niche, analyse the feedback and create a new competing App incorporating the comments that existing users have requested. You don't need AppTism to do this, it's just a useful tool. The better you know your competition (if you have any), the better you know your business.

8 Promotions

Promotions have always been an important sales and marketing tool, and that doesn't change just because it's a new emerging digital environment. You may have a critical time of the year, where people in your niche make lots of purchasing decisions, such as diets which are massively popular following Christmas and New Year.

With that in mind, you may choose a marketing campaign in January in an attempt to get as many new customers as possible. By doing this you have the potential to gain massive market share in a very short period of time. If you are not really competing with direct rivals, then you have the opportunity to whizz up the overall App charts with a good marketing promotion, carried out at the optimum time of year, month, or even week. You could consider competitions, price drops, bonuses etc. If you are interested in contests, then Apple provides further guidance.

> ❝ Most marketing goals fall into one
> of three categories ❞

9 Branding

By investing some time in marketing, you will be able to enhance your branding. Its your branding that will carry through all forms of your marketing, from emails, to App copy, to Screen-shots, to App graphics etc. The more targeted your App is, the more focussed on your ideal users you become, the better your marketing and sales will perform.

Your marketing goals

Do you remember, near the start of this journey, you wrote down what your App goals were. Do you still have them, have you looked at them at any stage during this journey? If you have those available, take a look at them now because your Marketing goals will probably be closely linked to your overall App goals. Most marketing goals fall into one of three categories-

1 Increase App sales on the App Store

Define your goal, and the Marketing required will present itself. The marketing campaign required to generate an extra 1000 App downloads per week will be very different to one designed to generate an extra 100,000 downloads per week.

2 Direct traffic to Apps product website

There are many reasons why this may be your preferred route-to-market. As the website is controlled by you, you get to control and measure every element of it. You might use a Squeeze page to capture their details and announce further App developments. You could use a survey to gain more information about your potential customers. You could show them a video of your App in action which may be more compelling than the 2D screen shots shown in the App Store.

3 Establish and promote existing services / products

Your App may be a vehicle to promote other products and services which is a perfectly valid use for an App. If this is your purpose, then everything about your communication should endorse and support the Product/service in question. This includes graphics, tone-of-voice, copy-writing, promotions etc. So identify what your Marketing goals are, and then stick to them with all your Marketing communication.

2 ways to your App

There are two main ways for your potential customer to get to your App.

Via the App Store itself: App Store browsing, Keywords, which specify when you submit the application to Apple, and Power Search usually carried out by a potential customer looking for a specific solution, App Store highlighted listings such as *Staff picks*, or the *New and noteworthy* category. The other main route is...

Via your own web page: Via internet search using keywords. You can attach these keywords to your web pages. You can of course direct any Advertising, Press releases, Blog posts, or Reviews to your web page. Very importantly you can also direct them straight to your App inside the App Store itself, by using a direct URL link which you obtain through the App Store. When a potential customer clicks on such a link, their iTunes will automatically be opened, and they will be taken directly to the appropriate App Store sales page.

Direct access

The ability to scoop up a potential customer wherever they are in the world, what ever the time of day, whatever they are doing and dropping them not only into your shop but next to the till, is incredibly powerful. The only action the prospect needs to take is to *tap* on one button, and the purchase is complete.

So unlike Jeffrey Hughes who wrote *iPhone and iPad Apps Marketing*, who implies that both routes are equally valuable, I suggest where possible you send them direct to the

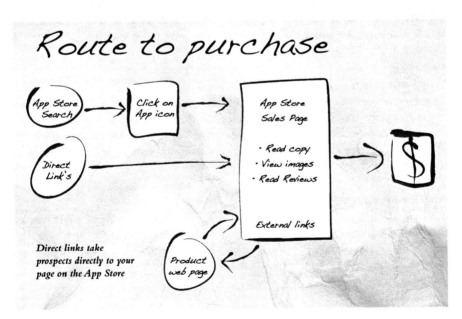

App Store. Whilst I accept that their is no harm in also having a web page to promote the App, I prefer to send people direct to the App itself, because that is the only place they can make a purchase, and in the great majority of cases where they will be searching for Apps in the first place.

However, I do still recommend that you have at least a simple web product page to support the application. One strong reason to do so is because at the time of writing there is no way to attach video to your App Store's product page. Until that changes you are forced to mount any such videos elsewhere on the web. If you produce a video, and I recommend you do, then place this on YouTube and any other video Social media sites for that matter, and link it directly to your App Store sales page for better sales conversions.

Communication channels

In our highly digital age the number of viable communication channels is daunting, but it's also very exciting. There exists today more opportunities for communication that at any other time in our history, and for the first time many of these channels are free to access, and in many cases free to distribute.

Here is a list of significant communication channels you can use to increase your App sales; Word-of-mouth, Trade Press, Web banner adverts, Internet search engines, Joint ventures, Co App advertising agreements, PPC campaigns such as Google's Adwords, App PPC campaigns, iAds from Apple, Viral marketing, On-line groups, On-line fan pages, Video websites such as YouTube, Network websites such as Facebook and Linkedin, PodCasts, Radio, Newspaper, TV, Email campaigns, Direct marketing, Direct links, App sales copy, Screen shot images, and App review magazines and web sites.

Your App sales will increase from activity in any of these areas, the magic question is which should you be looking at first, where would you invest your time? Here are the ones to concentrate on;

Word-of-mouth
To get great word-of-mouth recommendations you need to create a great application in the first place. Get a great developer to ensure that you get great App graphics and that all the App Secrets are included to maximise your App potential. Techniques to get great user reviews, which we have discussed in earlier sessions, will definitely help the popularity of your App. *Extra cost: None!*

App Review sites
There is a technique to getting your App reviewed by the many App review sites in the world. We provide this list to all our clients and explain how they can deliver that App to the review site so that they do not have to buy it to review it. No App review website is going to buy your application. *Extra cost: None!*

App Pay Per Click campaign

This is the only 'paid for' advertising mechanism I have short-listed, and I want to take a little bit of time to explain to you the benefits of this. This territory is growing fast, currently estimated at $68 million, it is estimated to rise to $1.3 BILLION by 2013. At present there are 3 main players in this territory,

AdMob – owned by Google, and MobClix: You can start an advertising campaign for as little as $50, your adverts will appear on Apps and websites, but essentially only for people browsing or using iPad, iPhone or iPod Touch users that is all the users who can realistically buy your application. You can also select geographical location, and other key advertising metrics.

This technique is highly measurable, so the trick here is start with a modest budget and tinker with the campaign until it is generating positive return-on-investment. Only once that is achieved should you ramp up your advertising budget, monitoring all the way to ensure you maximise your income. There is even a *free* option where you agree to advertise other people's Apps within yours, and in exchange they will advertise yours. Like an anonymous Joint venture deal.

iAds – by Apple: iAds are a recent introduction from Apple, and as always they don't do it by half-measures. iAds offer benefits other vendors cannot. For example when a user launches an iAd from your App, after they have finished viewing the advert which can contain games, video, audio, in-fact they can be a full-blown application in their own-right, they are then returned to your App once finished.

All other forms of advertising will take them away from your application. Unlike the Apps (30%), Apple will hold onto 40% of advertising revenue that your App generates.

Impulse purchases

Many Apps are bought on impulse. A great example of other impulse purchase products are those items you get in the Supermarket that are positioned at the end of the aisles by the tills. For those items to work effectively they must have bright dramatic colours to attract the eye, a produce name that is self-explanatory and enticing, and be at a reasonable price-point so that the customer can decide to buy without a great down-side.

It is probably no coincidence, that over 90% of Apps are sold for under $10, and although the average cost of Apps is rising because iPad Apps tend to hold a higher price point, many applications are still bought as impulse purchases.

As such your customers may not read every piece of copy, they may not examine every image, if it is related to an area they have some interest in, and not too expensive, on many occasions they will choose to buy, and evaluate the purchase later.

Some will become great purchases that they use time-after-time, some wont. There is little down-side because the purchase price in so many cases is so low. We take a look at the name of your App in a later chapter, here we will take a look at the App icon.

The complete Secret App formula

12 easy steps to App success

DVD 1:
Avoid Rookie Mistakes...

DVD 2:
Find Your App-ortunity...

DVD 3:
Test Your App Concept...

DVD 4:
Design Principals pt1...

DVD 5:
Design Principals pt2...

DVD 6:
Know Your User...

DVD 7:
Refine...

DVD 8:
Online Marketing...

DVD 9:
Social Media for Apps...

DVD 10:
Blast Marketing...

DVD 11:
PR for Apps...

DVD 12:
Optimize Ongoing Sales...

Visit: www.AppManSecrets.com/Shop12DVD_set.html

App icon

With a piece of text, the brain has to translate the shape of the letters into words, then attach meaning, and comprehend the message. The process is much more immediate with images that's why Road Signs around the world are image and icon based, rather than primarily relying on words and sentences.

So it's important to get the right icon. The right icon will quickly communicate what your App does faster than words can ever achieve. Don't try and be too clever, somebody browsing through the App Store will glance their eye across an App logo for just a tiny fraction of a second, so there is no time to be ambiguous.

❝ The process is much more immediate with images that's why Road Signs around the world are image and icon based ❞

Studies of supermarket shoppers found that consumers could recognize their favourite brands within a tenth of a second, that is how quick a brand has to communicate to be really effective. So if you have a fitness App, make sure you have an image that shows that, if you have a travel App then you need an image representing travel, *you get the idea.*

If you have a game App unconnected to other previous games, then you have more artistic license because there will be little you can do in the few pixels available to illustrate your game, but you can still use colours and images to indicate who you are appealing to. Think about the target audience as always, and that will direct your decisions regarding colour, style of images, and certainly the App name – which we will explore in much greater details later in the book.

You can of course also include text within the icon itself. And whilst the brain will process the image faster, the text can be used to confirm to the user exactly what the application is about.

If you have an App that shows a picture of a book, it may not be clear what the App is specifically regarding, but if the icon also says the word *Spell* or *Dictionary*, then it will be very clear to the viewer what exactly this application deals with.

If you are using a designer or developer to create your App icon, and have completed any of the exercises earlier in this book, particularly regarding your ideal user, then please do share that information with them, it will help them greatly. As a consequence they will be in a much better position to provide you with the right solution for your needs, and more importantly will be focussed correctly on your target customer, saving time and money in process, and with a significantly better solution.

Offline marketing

Offline media such as Radio, Newspapers, Magazines and TV to name just a few are often neglected as valuable forms of distributing information on your commercial products, in our highly Digitally connected lives. However don't be too quick to turn your back on this fertile, prosperous and proven commercial environment. Instead, let's see if you can make offline marketing work for you.

Radio

Few people ever consider Radio as a viable form of advertising or communication. However, with the increase of digital radio stations, as well as web based radio stations, there is a huge abundance of options. Whilst it may be very difficult to shoe-horn your App into one of the big commercial radio stations, it can be relatively easy with a smaller, niche radio show.

Radio shows can be surprisingly fun to do, with a DJ or presenter asking you questions, there really isn't much need to prepare. Within half-an-hour-or-so you have introduced your products to the radio shows entire audience, and if you ask them nicely you can even get a recording for your own reference and marketing benefit. You could add this to your website to help advertise your App further.

Some Radio shows will have a daily or weekly technology feature. What better way for them to not only mention your new App, but actually have the Creator *Live* on the air. It's great for them, it's great for you, and the listeners love it. Every one's a winner!

As usual the best way to find a suitable radio show is probably via the internet, search any details regarding the shows, or the DJs biography always looking for synergy between the subject of your application and their audience. Then fire over an email, or better still give them a call. Because you don't have the power to actually show visual images of your App over the Radio, don't expect huge sales traffic. However, any form of publicity like this is great – and you can then add a bit of text to your App description – *As featured on the X Radio Show!*

> ❝ it's never been easier to get your face, or your App, on TV than it is now! ❞

TV

Would you consider approaching a TV station to feature your App for Free? I suspect not. However you may have noticed that TV has changed, there are more than 15,000 Satellite channels, beamed from 420 Satellites, it's never been easier to get your face, or your App, on TV than it is right now.

Some of these stations will specialize in the same niche as yours, many specialize in technology and innovation which are therefore potentially interested in any application irrelevant of it's market niche. Even none technically aligned TV channels will often have a regular weekly show that deals with suitable genre.

If you're really ambitious, you can take a look at one of the National channels. Let's be ambitious...

The BBC is oldest, and most famous TV channel in the UK, it has two terrestrial TV channels and several digital ones, including BBC World which is exported around the world to over 264 million viewers and that doesn't include UK figures.

One of it's regular programs is called *Click*, a short but informative technology news program revealing the latest gadgets, and techno info web or Mobile based. If you have a stand-out App then their is a real chance they would feature yours. There's nothing to lose, and it will cost you absolutely nothing to approach them.

Newspapers

For now at least the newspaper is still alive, and still powerful, with huge daily readerships. All newspapers these days are backed up with significant internet versions, so your articles have the chance for a double exposure. It's arguably easier now to reach the appropriate Editors, and Journalists than it's ever been. Most will give their direct email address at the bottom of their column.

Consider which element or column your story will appeal to, for example;

> Game Apps – technology, or entertainment section
>
> Lifestyle Apps – lifestyle, entertainment, or technology section
>
> Health Apps – health & fitness, leisure, or technology section,
>
> *you get the idea*

❝ However any good Editor has one thing in mind – their Audience ❞

If you forward your information to the wrong person they may forward it to the correct department, but doing a few seconds research it's usually easy to establish which column or section you should approach.

I was an Editor and Assistant Publisher in my previous life with clients such as Virgin, Thistle Hotels, RoyalSunAlliance, EasyJet and many others. Amongst other things I was in-charge of producing their magazines from cover-to-cover. So I can tell you from experience that if somebody approached me with a suitable story to be included inside the magazine, where I didn't have to find and commission Journalists, or commission Photographers, or pay photo libraries for images, that was available imme-diately, and for free, then I would jump at the chance and so would any Editor.

However, any good Editor has one thing in mind – their Audience.

So like any publisher we were approached many times, with many stories and many articles all hoping for free coverage. The truth is only a handful made it through, here's why... Most of the Articles I received were focussed on their needs, not mine. If you want your Article featured, then you need to think about the needs of the Editor or Journalist that you are sending the article too, otherwise it is a waste of time.

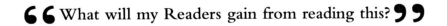

❝ What will my Readers gain from reading this? ❞

So here's a list of their needs;

Appropriate -

An article that is of genuine interest to their Readership, which is closely related to the Subject and Personality of the Magazine.

Fact based-

Interesting facts, figures, insights, or discoveries that make the Article a compelling read. This could just be a true-life story, however the more facts that can be established the more believable it will become.

Timely-

Is it a timely piece, perhaps specific to that time of the year, a particular event, or something that is a currently news-worthy

Readers follow-up

The readers have an option to get further information on this area if they are interested.

Available now

An article that is already-to-go, can be incredibly appealing. Many a PR article has made it into Print because another Story didn't come together in-time sometimes you just need to be in the right place at the right time. So if your story is sitting on their desk, and it's appropriate to their audience when their own Story runs into trouble then that could be the opportunity you have been waiting for!

In time

It doesn't matter how good your story or article is if they have already gone to print! At the very best you will have to wait until the next issue, at worst you have missed your opportunity. Ensure that your Story arrives at the right time. To do that it's easy, just call them.

Look in the magazine, find the telephone number for the Editor, they aren't too hard to get find. Speak to them directly or their assistant and explain that you're sending through an Article on *X*, which you feel would be right for their audience because of *XYZ*. When is their next print slot, and how long before that would they like the piece?

> **❝ Ensure that your Story arrives at the right time. To do that it's easy, just call them ❞**

It's easy

And by doing that, you can actually test your article's content even before you write it so you don't run any risk of wasting your valuable time. If you want to get into a weekly local paper, they probably only want to receive your article a few days before print. If its a monthly, then they may need it up to two months before release date. And REMEMBER, a March magazine comes out in early February.

So if you want an article in the March edition of a National or International Monthly magazine, then they might need to receive your article by mid-December!

Plan

The potential of all of these areas are great. However, because there are no guarantees, few even attempt it, and that's an over-sight. All of these will take a little time, so here is my recommended strategy;

> **Get a list of sources**: All of which are easily available

> **Create a short-list:** Only select those that offer a close synergy between their Audience and your App, both in-terms of subject matter, but also approach.

So if you have a Finance App that helps Mothers save money from a tight household budget, then don't short-list Financial programs that examine Stock Markets and Investment strategies. Make-sure you stay in your Niche. That's where you are strongest and

Measure

It is impossible to track conversions inside the App Store, however if you direct TV viewers or newspaper readers to your website, you can do that using a traceable link, so you can see where your customers have come from.

Or you could even ask them through the App itself!

where you will get the biggest interest, and be able to build relationships and financial rewards.

Contact: Where-ever possible contact them via phone. You don't need to go into Sales mode, just explain what your Story is, what the Synergy is between your Story and their audience, and ask how they would like to receive the information.

Send it: Send it promptly, before they forget they had the conversation.

Follow-up: then re-contact them, just to check their received the article, and re-iterate how they can contact you if they have any Questions, or would like further information.

❝ Everybody has a story. What's yours? ❞

Remember

The Editor, Journalist, Reporter or TV Producer only cares about how the viewer or reader will benefit from reading your story, or seeing your piece. So think about the story around the App, rather than necessarily about the App directly, such as-

Granny wanted to find way to play with her Grand-children, so she created an application

Or, Mother wanted a Holiday, so developed an App to help her keep a track of her expenses and save for the Holiday she always wanted. Three months after the App was launched she had earned enough money to fly the family first class to Australia. Everybody has a story, *whats yours?*

So remember you are delivering them a free story, one that is going to appeal directly to their audience, and that you are assisting them by providing this knowledge. Don't have the attitude of asking for a favour, or some free advertising, that isn't a winning approach. Always make sure that they can get in contact with you, via several forms of communication for further details and information, if they need to do so.

App copy

App name

Here is a diagram we looked at earlier in
the sessions.

The first thing that anybody will see of
your App inside the App Store will be the
icon and App name. Therefore it's quite
obvious that you need to make those as
appealing to your ideal customer as you
can, so that they *click* on your logo to
discover and at least consider a purchase.

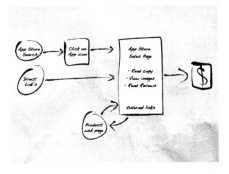

If they don't *click* they can't buy it's as simple as that. So it's worth us spending a little
time to look at what makes a great App name. First point here is-

Don't make your Audience work

Like you your audience has a lot of demands on their time, and they are used to being
bombarded by information, every second of the day. As such, by-and-large, they want
easy routes. Generally speaking they are not interested in high-brow App names that
'make-them-think'. The very last thing they want to do in-fact is think.

> ❝ If they don't click – they cant buy –
> it's as simple as that ❞

So if they are looking for an App to help them with currency exchange, calling it
World Trade App isn't going to get them. Call it *Currency convertor* or similar so they
know instantly that you have the App they are looking for.

It's very easy to come up with convoluted high-brow names that appeal to you, but
what your audience wants is clarity and simplicity. By-and-large App purchasers want
to find the App, download it and use it immediately. They are therefore looking for as
easy a path to that goal as they can find, which is why they used their iPhone or iPad
in the first place. It's the most convenient tool around today.

This is *not* the time to be using Greek mythology, or Latin references. Please do not
turn an incredible intuitive, easy-to-use, tool like the iPhone into a complicated hard
to understand device. Your customers will not appreciate it.

What's the function

If your App performs a core function, such as the *Currency convertor* example, then wherever possible include that within the title. This is particularly true for non-game Apps. If you're producing an entertainment App, then talk about the subject matter so that if it's talking or showing illusions, use related Keywords such as Magic, or Magician. If it's talking Hollywood Films use Film or/and Review in the App title.

If it's a full-on game then you tend to have more flexibility, and the App name will typically be about the characters within it, like *Angry Birds* or describing or hinting at the type of game it is, such as *Need for speed* which is a car racing game. Don't use convoluted long words. Keep it simple, and accessible to everyone.

Short

Although it's useful to have a short name so that people can remember it, it's essential if you do not want your App name to become truncated or abbreviated inside the App Store. If your an iPhone owner, then you know that there isn't much room for the App name. As a rough guide there is room for approximately 12 characters, however the exact character count depends on which letters you use, as every letter has a different width.

So keep your App name to within 12 characters and you will avoid problems, this can be a challenge though, and you may need a few attempts to find the best solution.

Here's some App names

Create a Car	iFitness
AA Route Planner	UK Train Times
RingTone Designer Pro	

None of these are particularly creative, or even impressive, but their sales are, so make sure your App name reveals what it does where possible.

Serial Apps

If you intend to create a suite of Apps, around a central theme, then you may choose a name that has that flexibility, such as;

App Man – Secrets	App Man – Fav Apps
App Man – App Reviews	App Man – App Ideas

If you don't have any plans for more than one App yet, then just create the best App name you can now and worry about that situation when the time comes.

Domain availability

Is your proposed App name available as a domain? In most cases you can tweak the domain name to suit your App even if that domain is taken. So, if you create an App called Granny Racing, and for some bazaar reason that domain is already taken , then you can adjust to www.GrannyRacing App.com for example.

To do this simply go to one of the many Domain registrars out there. One of the biggest is GoDaddy, or I prefer www.Names.co.uk. Type in the name there and see what the availability is like.

Keywords

In an earlier exercise you made a note of

Did you know...

Angry Birds which is a massively successful application is currently being considering by Hollywood as a feature film, just shows you the power that Apps can have.

Would you like to see your App made into a Hollywood Film?

your Keywords, you can use those here. Or feel free to pay Google another visit, and use their Keyword Search Tool, to obtain the most popular keywords for your specific niche.

Simply selecting the keyword with the top hits may not be the correct approach. Always consider the needs, desires, and thoughts of your ideal customer and let that knowledge direct your decisions. So you've got the basic idea. Please keep these principals in-mind as I take you through my App Name exercise...

App name exercises

Complete the following App name exercises, and you will quickly discover what the best name is for your unique situation, and your particular niche. If you are still not confident you have a name you are happy with then test it. Ask friends, colleagues, ask people via social media.

When you get feedback, just remember to keep your ideal customer in mind. If they are not your ideal customer, their opinion is of course still valid, and it's worth considering but if you think their suggestions aren't right for your niche, you are probably right. Stick to your instincts. Nobody knows your App better than you.

App sales letter

Your App will live on the App Store. Within reason you can add any relevant text or images, to help publicise your application. However, there is no way to adjust the layout, colours, or background to the page itself.

The key features are;

App icon

Price

Description

Screen-shots: These are very important, as you can see on this page, it's the images that attract the eye, rather than the text. This one reason why having great visuals is so important to a great App which is why we highlighted that in previous Sessions.

Customer ratings

Formatting
There is no way to adjust fonts or format the text in any way. The only way to do any type of formatting is artificially using punctuation marks to give emphasis to certain elements of text.

Less is more
You have a maximum of 4000 characters, but there is no need to use them all. Keep descriptions tight, informative and descriptive. Again, avoid being ambiguous. Follow the suggested format described soon.

App name
- What does your App do?

- What are the customer benefits?

- What are the unique features?

App name

- List the 5 main keywords in your niche

-------------------------- --------------------------
-------------------------- --------------------------

- Write down 10 potential App names

-------------------------- --------------------------
-------------------------- --------------------------
-------------------------- --------------------------
-------------------------- --------------------------
-------------------------- --------------------------

- Cross-out any over 12 characters long

- Cross-out any names that are ambiguous

Write down your favourite here...

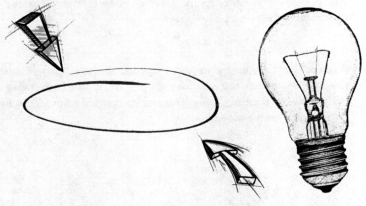

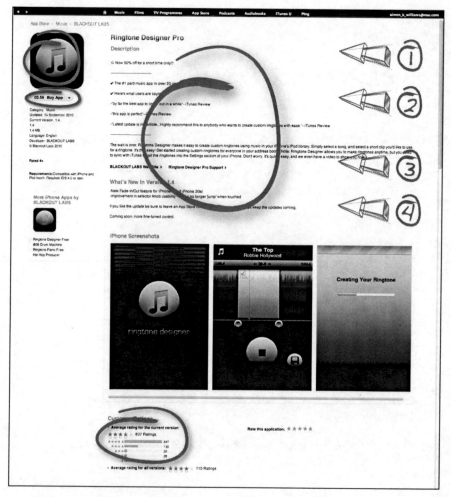

1 Promotion, why buy now

Can you give the viewer an incentive to make that purchase decision now, rather than delay purchase until another visit?

2 Testimonials

If you are just launching your App, you wont have any App testimonials yet. If you have any quotes that you have shown the App to from friends, consider using those for the time being. This can be updated easily in the future after getting some great reviews.

3 Emotive description

This is the place to actually describe in emotional terms what the App does. Consider using the storyboard approach so that users know how and when they are likely to use the App. Highlight the benefits rather than the functionality.

So if you completed the exercises earlier you can use those notes here. You will also find the comments you made regarding your ideal user of great benefit. If you wish to list features, do so in bullet-point form. *Last but not least...*

4 Whats new

If you have improved the App in any way, do list them here. App Store users like to know that you are enhancing and improving the App. It also demonstrates to them that you care about the App.

Customers know that every piece of software contains at least some minor bugs. The fact that you have made an update to improve them illustrates to them you are making a commitment to iron out these as much as is humanly possible. Consequently they are going to have a great user-experience.

App Review Websites

The list of App Review sites are growing all the time, each one with a different submission policy. Generally speaking reviews are done impartially and for free (as they should be) however there are some App review sites that request payment for Reviews.

❝ Don't pay for App reviews ❞

It is your decision if you wish to consider these, however my recommendation is to avoid them. They are offering little by way of service to their readers, and as a result offering little by way of service to you. None of these websites are going to prepared to purchase your App to review it, so there is a cunning way to get around this.

Apple are kind enough to provide some digital coupons with every App released. You can pass on these credits one-at-a-time to App review sites so they can investigate your App thoroughly without having to make payment.

Reserve these for App review sites, don't be tempted to use them for other marketing, friends and family. Even if you don't use them all initially, you will have them later when a suitable promotion or event takes place.

Blast marketing

Blast Marketing is a principal first employed by one of the world's largest industries, the Entertainment industry. All the indications are that society is moving towards an ever-more emotionally engaging entertainment environment.

It's Entertainment that turned a struggling computer Company into today's largest technological Company in the world. Apple achieved that by providing Entertainment through music, mobile applications and film using their Apple TV device.

As such the entertainment industry is incredibly competitive. A film may easily spend *millions* of dollars to promote a film, and when you're spending that type of money, and competing with rivals who have that type of money to invest, you better spend every cent wisely or you squander millions without any pay-back.

So what do they do

They align all their Marketing efforts to one specific event, at one specific time. In the case of film industry – it's launch day.

When their marketing has been really successful, the momentum from all their activities across Radio, TV, billboard posters, leaflets, countless interviews, press-releases, news gossip etc. reach a critical mass, where it deemed *newsworthy* in it's own right.

If they are able to reach that point, they will literally get *millions* and *millions* of dollars worth in press, newspaper, and TV coverage, all of which comes for *free*, but only as a consequence of all their marketing initiatives before-hand. I am assuming you don't have millions of dollars to Market your App, but we can use similar tactics to maximise the impact of your promotions to rocket your App up the charts.

The App Store...

By aligning all your marketing initiatives to a specific event, and in the case of application, we want to use a specific date, then that activity will generate a *spike* of traffic and downloads. This *spike* may be artificial in the sense that the traffic could not be sustained at that level indefinitely, however this technique has the potential to *push* your App into the top 100 downloaded, and certainly into the top options, within it's App category.

You can place your application in two of the following categories; *Games, Entertainment, Utilities, Social Networking, Music, Productivity, Lifestyle, Reference, Travel, Sports, Navigation, Health & Fitness, News, Photography, Finance, Business, Education, Weather, Books, and Medical.*

Eye-balls

By *pushing* your App to the top, it gets *huge* exposure, because that's where everybody is looking. Get it there once, and if it's a great application that people like, and that serves it's niche well, then it has a great chance of staying there, or there-abouts with very little, or no further Marketing activity.

Blast Marketing maximises your time and energy, and any investment you may have placed in advertising the application. Pick your event wisely, it doesn't have to be launch day, you may prefer to do a *soft* launch, gain feedback, optimize the price point, listen to feedback, before doing a *hard* launch with more vigor, energy and possible investment.

If you have a fitness App, you may do a marketing push after the New Year when you know there is huge amount of interest in diets, health and fitness, which provides a particularly exciting opportunity for your applications sales. If you have ski App, you may launch it at the same time as an international ski competition such as the Winter Olympics, or when the demand for ski holidays is at it's highest.

By knowing your niche, and knowing your customers, it is easy to find an appropriate time to *push* your App. Of course nothing is stopping you doing another marketing *push* in the future, however *do* follow the same principals and select a particular event, even if that event is a temporary App price reduction, and *blast* your App up the Charts!

Here's a few pointers-
Focus on one event

Such as the day the App appears for sale in the App Store, remember you can specify a launch date with Apple, if you plan ahead.

Build anticipation

Release a few screen-shots to wet people's appetites. Put out a *Coming Soon* message via Social Media to build anticipation. Your Apps product web page could have a countdown timer, highlighting the soon to be released application. Send out an auto-responder to your database each day for a week prior to the launch, each time counting down the days until the App is available for purchase. Highlight any launch discounts you may have, again to encourage traffic on that day.

Focus on the Story

Don't just talk about the functionality of your App, get emotive. Rather than saying this 'App tells the weather', say 'This App will make sure you don't get caught in a rain-storm without a proper jacket, so you get to your meeting without looking like a drowned rat'. Focus on the benefits rather than the features.

Incentive for purchase

You must find a reason for people to put your App at the top of their *to do* list, or it will end up at the bottom of the list where it will remain. Must be time-sensitive. Perhaps your App will be a discounted rate for just one hour only.

Get where the eye-balls are

Using Blast marketing in conjunction with a Great App is your best way to get your App to the top of the Tree.

App-ortunity Marketing

Learn your goal   Market niche

Concept   Competition

Pre launch Marketing

PR Trade mags   App video

App web page   App name

Icon   Social media

App video   updates

App launch Marketing

App Review web sites

Purchase incentive

App PPC ?

Optimise price

Refine your App

If you follow the exercises contained in this book, you will have discovered the following;

- ✓ Your App goals
- ✓ 15 great App opportunities based on your own existing interests and knowledge
- ✓ Evaluated the commercial opportunity
- ✓ Examined the direct competition within the App Store
- ✓ Selected your best App opportunity
- ✓ Know your customer, what they are thinking, what their situation is

You will have even learned the five main App architectures, and probably have a pretty good idea whats going to work for your unique situation. You might even have a sketch or two of how the App architecture might flow.

If you have any of these points covered, then *congratulations* you are well on the way to constructing a highly commercially focussed application. If you haven't completed these exercises, this is your invitation to do so. You may alternatively trust your developer to complete all these exercises on your behalf. Do realise though that few developers offer such advanced features, and if they do they will charge accordingly and you will still have to carry out the same work, so it is time well invested.

We introduced you to the Rookie mistakes earlier in the book. Here is a great chance for us to elaborate on these points, refining your application concept into a fine-tuned commercial product. Your App will be able to stand shoulder-to-shoulder alongside experienced App creators that know all the tricks of the industry and ensure you can compete in your own right.

Apple have produced an amazing *route-to-market*, that is open to all. This book hopes to continue that approach, give the skills and insights from an industry insider to you, so that you can compete with the best. I think you deserve that opportunity, *don't you?* So here goes...

Build a list

If you have ever used internet marketing to drive traffic, produce leads and customers, you will know just how hard it is, how long it takes, and not least of all how much it costs to build a significant list.

Unfortunately there are many get rich quick schemes, many involving the so-called magical Google Adwords PPC marketing campaigns. However in my experience, you need to have a very fine tuned marketing funnel to build a list with any pace, and without blowing your entire marketing budget in one night.

> ❝ **Unfortunately there are many get rich quick schemes** ❞

You also need an established funnel, or *back end* which will take advantage of the traffic that comes in. Only then can the expense of the initial marketing be justified, and have any potential for financial return.

By contrast, with a popular App, you can easily have thousands of new customers per month, per week, per day, and in extreme cases even per hour, and unlike PPC campaigns, it wont cost you a cent!

Now I do realize that you might be completely happy with the direct profit your App is going to produce, but there is no reason not to be greedy here. It's an amazing opportunity that can have huge commercial prospects, so why not take full advantage.

So I strongly recommend you include a mechanism within the application to build a list. The easiest way to achieve this is by simply having a link from your App to a Squeeze page – the name given to a web page designed to obtain your name, email address for future marketing announcements.

To achieve this, you must place a compelling reason in your App, to entice your customer to *click* and then hand over their valuable contact details.

The most alluring compelling reason is of course different for every App, and every user, the one you select obviously needs to be right for your situation.

You may already have a product or service already available and perfect for this situation. If not consider an eBook, video, or a bonus level to the App if that fits.

Get great user reviews

Would you think having a good review in the App Store would be a good thing? You'd think so wouldn't you. So why is it so many applications don't have any mechanisms in them to achieve this?

Don't make the same mistake. You can't put a gun to your customers head and force them to give you a great review, but you can at least ask them, that's not so hard is it. So why not include a pop-up that asks your user to give your App a review. Now if you're clever, you instruct your developers to make sure this pop-up appears after 20 or so uses.

App users are a fickle bunch, and I include myself in that. Chances are if your customer doesn't like your App, then they will have given up on it long before 20 times, so the reviews you do get are largely going to be positive – and that's only going to help the adoption of your application.

The more great reviews you get, the more customers you get, and so on, and so on. Its a perfect cycle.

If you do get a negative comment, it gives you a great opportunity to address that issue before other customers become disgruntled.

Optimize your price point

Optimizing your price point is vastly under-appreciated. In my former life as a Product designer I worked for a Company called Historic Royal Palaces, which as the names suggests runs many of the UK's top tourist locations. In-fact if you have ever been to the Tower of London, then you will have seen some of my products.

As a designer it was always revealing to learn how commercially successful a new design had proved to be. We examined where it was positioned in the shops, in which Palaces, what lighting, but by far the most critical factor, perhaps not surprising is of course the price point.

You may suspect that if you raise the price a little, you generally get a few less sales, and that if you drop the price a bit, you get a few more sales, and at some point there is a magical optimized price point.

That is not my experience...

On many occasions we had a product we were excited about that produced almost no sales whatsoever. We would of course tinker with price points, and most often than not, at a particular price point the sales would rocket. Its amazing to experience, so rather than think of your price point being like a dimmer switch, think of it much more as On or Off switch.

❝ you don't need to learn complicated pricing theories ❞

Sales don't fluctuate gradually depending on price-point they either leap off the shelves, or drop off a cliff, so taking a little time and effort to find your perfect price point is time well spent.

There are countless philosophies and strategies with regards to price points. You may have, for example, realized that lots of internet marketers products end in the number seven. This can be accredited to Ted Nichols who first announced his theory in the 1980s, marketers have been proving his theory correct ever since.

This is just one of many theories. Nicholson, Elad and Stolarz explain their theory on App pricing in their book Starting an iPhone Application Business, which suggests a sensible look at the market, the competition, size of your niche etc. which all sounds very sensible, but that approach is also out-of-date.

Apple provide every developer a web based portal to control, and review your App status. This is the portal that controls your sales price, images, sales letter etc, all of which can be adjusted at any time. At the time of writing everything except your App's keywords could be adjusted with near immediate effect. As a consequence, you don't need to learn complicated pricing theories, instead I suggest you guess.

By using a simple spreadsheet you can measure the quantity of your App downloads, the price, and easily calculate your income. I suggest you adjust the price several times, each time adding the data to your spreadsheet.

In just a few days or weeks, you will be able to explore all your realistic price points, and easily see which is the most optimized price point for you.

You can then hypothesize as much as you wish, as long as you adjust the Theory to fit the price point, not as many do, adjust the price to fit the theory.

> **❝ Don't adjust the price to fit the Theory, instead adjust the Theory to fit the price ❞**

At that end of the day, only one thing matters, the price your customers choose to buy at. Don't adjust the price to fit the theory, instead adjust the theory to fit the price.

Help and Support pages

In an ideal world your App would be so entirely intuitive that there would be no need for any Help or Support pages. Apple themselves demonstrate this by not supplying them with their applications. But don't be fooled...

The reality is that Apple produce the core applications, which to some degree if you like them or not (and I'm sure do), you have to use them on a very regular basis, often every day. It is therefore only a matter of time before the user becomes equated with Apple's Apps, and they feel intuitive.

Users are a fickle bunch, and unlike Apple's Apps where they are forced to learn them to some degree, they wont afford you the same curtesy. The more original your navigation is, the more support pages you will need to provide.

One technique which often works well, and is worthy of consideration is the Help pop-up screen. You can elect to display just on the first time of using the App, or until they *click* a "Do not show again" button. It's very simple, but it works.

Consider using video, there is no better way to show how to use an App, than by showing it in real time through video clips.

You can use these same video clips to aid promotion through Social Media sites such as the ubiquitous YouTube, as well as include in any web sites you may use to publicize your App. You can even send it out as an email shot.

I promise you once you have planned your pages and built your App flow, that you will feel confident that everybody will know instinctively how to operate it, but this is unlikely to be true, so here's a warning...

There is little in life that can be more frustrating than handing over the App that your development team have sweated over, only to see your test-user completely get lost, *tap* the wrong buttons, and not even get to grips with the most basic of functions. It is a humbling experience.

This is not the time to panic, this is the time to get the insights, and learn from your potential users. Why did they touch that small button there, not the big throbbing green icon at the top? Why did they not think to swipe the screen when they swiped all the pages before that one?

Human behaviour is a wonderfully chaotic system and is a valuable branch of science in its own right. As an App creator it is unlikely that you are going to become human behavioural expert, so instead take the approach of all feedback is good feedback.

Remember that any insights gained at this stage, prior to release, are truly golden. Enabling you to make significant improvements, enhancements and modifications prior to commercial release, you are saving months, and development costs by learning all you can in advance of launch.

Great feedback loops

If you carried out the earlier *Know Your User* exercises, then you will have had a little experience of how getting to know your user is vital for providing amazing products and services to fit their needs. It stands to reason.

The more you understand your user the better your products will become, and the more money you will make.

How many App reviews have you given?

As you might imagine, I have downloaded lots and lots of applications, yet at the time of writing I have yet to submit even one review. *Have you?*

So it's clear you shouldn't expect to receive hundreds of reviews either. So you need to find a better way illicit the types of user information that can improve your App. I call these techniques Feedback loops. Any technique or method that results in user responses and feedback is good. You can consider incentive schemes, such as a freebie when you receive information, or they complete a survey. You could include them in a prize raffle, what ever you think will work.

The alternative is to incorporate extensive analytic code that will record data from your users behaviour. If your are already familiar with Google Analytics, then you understand the basic principle, here's how it works...

Your developers can incorporate special bits of code into your App that will track their behaviour and send that data back to you for analysis.

66 Apple use a classic product funnel 99

If you have ever been asked by an App to reveal your location, even when it seems irrelevant to the App your using, chances are the App is attempting to record your location for exactly this purpose, and nothing else.

Analytics can be used to record, which Country your users are in, what language they are reading in (assuming the App has various language options), how often the App is launched, how long it is used for, which pages are sot frequented, which pages are used, and so on, and so on, *you get the idea*.

This data is then sent electronically back to you for analysis, without the users knowledge, so that you can use these insights to drive enhancements.

When done well, not only will you be able to make improvements evolving your App into a refined and Competition leading application, it will also give you the insights and direction to produce other products and services that fit your audience, which brings me rather nicely to...

Up-sell

Aligning another product or service to your application can have massive impact on your revenue income. This cannot be stressed enough.

Lets learn from the best

Apple lure you in with iTunes their free music playing software. You may then choose to buy a few songs, a couple of albums each costing a few dollars. Then you may be tempted into buying an iPod, then an iPhone, then an iPad, then the all singing and dancing Apple desktop computers, and perhaps finally enrolling into an Apple training course. This is a classic product and marketing funnel, or up-sell from one product to another, and it works.

iTunes	Songs / Apps	iPod / iPhone	iPad	iMac
$ FREE	$ 0-30	$49-299	$499-829	$999-4999

Illustration shows how Apple's products are strategically arranged from the cheapest product, such as iTunes which is free, to their most expensive desktop and Mac Pro computers which range up to $4999.

However the quantity sold of each product will be inversely proportion to the cost, that is Apple's has billions of iTunes and App customers, with much fewer customers purchasing their high-end machines. This is one of the biggest reasons that Apple invented the iPad, otherwise there would have a been too big a void between the iPhone and iPod to their desktop and Mac Pro machines

If Apple only had their high-end desktop computers they would not be the biggest Technology Company in the world, which they are today. Their success has largely been the creation of lower priced consumable products that have expanded their customer reach, their profile, and rather significantly their income.

People forget that just a few short years before the iPod emerged as one of the coolest electronic products of all time, Apple was on the verge of bankruptcy and real threat of extinction or take-over.

So learn from the best, and consider what other products and services you can incorporate into a product funnel.

Apps provide an amazing low-cost or *free* entrance to join your product funnel for your customers. Where you take them from there, is your choice. You may have products and services already suitable, if not then in-time these can be created, and what better way to do that than by listening to your customers and fulfilling their requests. They will be delighted, you have listened to them, and happily pay you for the privilege. Lets take a look at exactly how this can work for Apps...

Up sell example

Let's use *Michael* as an example. Michael is a keen fishermen, and decides to create an application to help other enthusiasts like himself. The App details where are the best lakes and rivers to catch different types of fish, as well as giving baiting tips, and equipment advice. He names his first App *iFish*, which he sells at $1.79. After proving successful, he creates another application following some user feedback, which includes additional features, such as allowing the user to photograph and log all their catches, where they caught them etc. Users can pass on tips, and get tips from others using application. He calls this enhanced application *iFish PRO*, which he sells at $4.79.

The final step in his product funnel, which is an off-line fishing event where entrants fish against each other in a days fishing event, in hope of proving themselves as the best fisherman of the contest, is called iFish Challenge. Michael sells this event for $97.00.

Applications such as Tap Tap, received over 1 million downloads within two weeks! For this example however we will use very conservative figures.

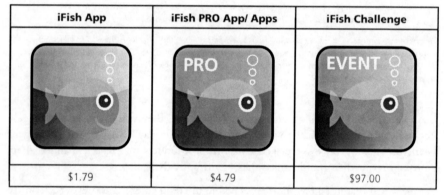

iFish App	iFish PRO App/ Apps	iFish Challenge
$1.79	$4.79	$97.00

Placing your application at the start of a product funnel, is the right way to maximise your commercial opportunity.

Michael's initial App achieves only modest success, receiving approximately 50,000 downloads per month, after Apple has taken their 30 per cent this leaves Michael with $ 62,650.00 profit (minus initial App development costs). However due to his other two products in his product funnel, this profit grows to a much more enticing $ 172,462.50 in his first month. Even allowing for no improvement in sales throughout the year, Michael will make $2,069,550.00 profit (minus original App development costs and any associated event costs).

Product	Units sold	Profit
iFish	50,000	$ 62,650.00
iFish PRO	25%	$ 41,912.50
iFish Challenge	2%	$ 67,900.00
Total monthly profit:		$ 172,462.50
Total yearly profit:		**$ 2,069,550.00**

Apple's costs have been factored in. Only costs of initial App development, and any associated costs to hold the event need to be accounted for.

By comparison a popular application like *iShoot* which made $600,000 profit in a month, would receive profit of $ 15,813,000.00 if it were placed into a similar product funnel working on the same percentages.

Take action

I personally believe that everybody has a million dollar App inside them, my task, is to help you, find yours.

Many people get very excited when they tell me how they had a great idea for an App ages ago, and now somebody has created the App and it's been a huge success. When this happens they are always grinning from ear-to-ear, as if they have won some kind of prize, but what have they achieved – nothing.

❝ Ideas in jars, don't serve anyone ❞

At worst they had an opportunity to provide a valuable experience, asset or knowledge to others and elected not to do so. Where is the achievement in that? Is that worse than not having the opportunity at all?

It's absolutely right that it is the person that brought that idea to life. Or to put it a better way, the person or Company that delivered that asset to the people that wanted it, that gets financially rewarded. Ideas in jars, don't serve anyone.

So if you do have a good opportunity, I *urge* to explore it, and if viable, produce and deliver that asset to the Universe, and make the world we live just a little bit better as a result, isn't that the cornerstone of social development?

This book has been written with that in mind. I think it's only reasonable that somebody new to Apps may want to do some homework before they commit to an exciting project like an App. I want to deliver the facts, the insights and Secrets from within the industry so that you have all the knowledge and tools available to deliver your opportunity, and your asset to the public, I am very committed to achieving this which is why I speak in various Countries, run Web based seminars, live conferences, Corporate consultancy and more.

I do not want to look at myself in the future knowing I had the opportunity that I did nothing with. Ultimately though it all comes down to one person, and that person is you, and that's just how it is, the decisions you make define who you are, and how you are seen by others. May I share a personal story...

Great position

You are in a very fortunate position. You don't need to create your own Company just to get your App produced. If you choose to, you can take immediate action and have your App in the App Store in a couple of months. Or alternatively you can watch

My story

My first ever App was to be a map of the London Underground. I knew from experience that I was always looking at tube maps, and attempting to estimate the journey time to get to my next meeting, and changing routes at the last minute – it was always a hassle.

So I had the idea of putting an interactive map in the hands of London commuters, so they would know exactly the shortest journey, when to leave to get to that important meeting on time, or just how long it was going to be until they got home that night.

You could select your route by tapping a start and finish point on the map, or select the stations from a list, or use the keyboard to type them in. It showed the duration, the fastest route, and was easily editable if you needed to tweak the route. To reset – all you needed to do was shake the iPhone, and you could start a new journey.

However as a graphic and product designer at the time I had and no software development experience, so I attempted to put a suitable team together in the UK. This was the early days of the App Store, what I found was there simply wasn't the skills I needed in the UK at that time.

As a result I felt I had to build a team in India. All-in-all it took me nearly six months to put the right team together. Shortly after we started App development proper, guess what happened... Yep, a London Underground App was released.

That App is still one of the biggest selling Apps of all time with millions and millions and millions of downloads. Because of the delay, we had missed a golden opportunity to get our App to market first.

It was only because I adapted our application to the Moscow underground system, that my story became a successful one, without which you wouldn't be reading this book right now. And I would be reminiscing on the day that I missed my million $ unique App opportunity.

If you have an opportunity, take it, or somebody else will!

somebody else get rich from your opportunity. So although I applaud you for researching your industry by reading this book, it will be your decision to take action or not that will ultimately decide your future and what you deliver to the world.

Take action

As you have journeyed through this book, you will have discovered a great deal about your personal opportunities, and my hope is that if you get nothing else, that you get how accessible this opportunity is to you.

As you near the completion of this book, you are likely to be in one of three very different positions. For some of you this is not so much the end of the book, but the start of a new and very exciting new chapter in your life, let's call it your *App-ortunity*. For other readers this is the *end*, and will return to their regular lives, living day-to-day as if nothing had changed, and is if no learning had ever taken place.

If you are the later, then I am sorry this book has failed you, and I wish you well in whatever it is you do, and there is little in the rest of this book that is of relevance to you, so you may as well turn on the TV instead and save yourself some time.

For those that wish to take action, this is where the excitement starts. There are few things more exciting than launching your very own products and services around the world, it is a truly growing experience. Or as one of our readers put it *"It's better than sex actually"* Sam Brown, although we can't promise that.

The guidance and knowledge you already hold as a consequence of reading this book already places you in a very high category, one that few other Rookie's have the advantage of. Your very first application will have the forethought, and knowledge that only a seasoned pro developer can usually obtain, and you have accomplished that by simply investing some time and energy into reading these pages. That simply fact alone, separates the men from the boys in the world of business.

As a curtesy to you, because I don't just want you to take your leap into the App Store without the support that you need, I offer two suggestions...

Suggestion 1
If your App concept is still un-defined, this is for you...
www.AppManSecrets.com/ShopGold.html

You may have enjoyed the book, you may have even identified various App-ortunities but may have found it hard to select them, this is the right options, for you. Alternatively you may have skipped the odd chapter here and there, or maybe even rushed to the back of the book to search for the punchline, and perhaps have not completed all the exercises then this is definitely the right option for you...

We have put together an amazing interactive App creation course called *The Secret App Formula*. Over 12 sessions, this internet based seminar course will teach you exactly what the best App-ortunity is for you, how to develop it, how to create revenue from it, how to get development costs and timing schedules, and how to launch your App into the App Store with a bang. You want to be successful from day one – *right!* Here are just a few of the comments received from participants...

> *"I feel truly blessed to have the opportunity to have you to guide me to develop such an application...*
> *"WOW.......... I CANT WAIT TO COMPLETE MY APPS AND GET IT LAUNCHED"*
> Dr Nor Ashikin Mokhtar

> *"The seminar gave a great insight into the possibilities of Apps. Simon was a genuine guy and shared a lot during his talk. I am sure his course will provide a shortcut to success"*
> Rebecca Guppy

> *"Sessions 5 was awesome!"*
> Raymond Cheong Choong Ee

> *"I really like Session 6.... I have already done a prototype on Excel and the session 6 just completely helped me firm up the loose pieces... It has also helped me write up the intro page on the APP store.... The part where you link up the emotion decision while purchasing the App helps a lot. Thanks.*
> Bill Sai Hua Koon

> *"Dear Simon, You are a great trainer, teacher and mentor to us in this program. Your willingness, transparency in your sharing and especially your patience with us. This is really a great program that I will recommend my friends to sign up. :-) Cheers!"*
> Sharon Ong

You will receive also the comprehensive 12 DVD pack, so that you can go back over any elements as you wish, and will have that valuable reference aid at your disposal at all times. This is great for your first App, but invaluable for App two, three and more.

You are welcome to enter our next course at *www.AppManSecrets.com/ShopGold.html*, it will be a pleasure to see you there.

If however you have been following this book religiously, completing all the exercises, and as a result have a clearly defined application concept, and need to find out how to take it forward, then you need suggestion two...

Suggestion 2
If you already have already identified a great App-ortunity, this is for you...
www.AppManSecrets.com/RichAppPoorApp.html

If you have some or all the exercises completed, either filling out the book as you went along or making separate notes and references, then this is the right suggestion for you.

We call it the App Inspection, it is a comprehensive report and investigates every commercial aspect of your concept.

Simply visit the book bonus page listed below (lots of other good bonuses there too), and click on the relevant link to receive your App inspection report.
www.AppManSecrets.com/RichAppPoorApp.html

The App inspection report includes;

>   Commercial summary of the mobile App market

>   Technical details of the application

>   Elements that will be completed by the development team

>   List of elements that you will need to provide

>   Summary of the App proposal

>   App abstract

>   Story board of ideal user

>   Context for the Apps use

>   App characteristics based on the ideal user

>   Summary of potential App architecture

Full App assessment which includes;

>   App assessment review

>   Size of niche

>   App name

>   Clarity of task

>   Word-of-mouth potential

>   Revenue options

>   App introduction

>   Graphics

>   Navigation

>   User help

>   Price point

>   Opportunity to Build a database

>   Feedback loops

>   List of associated keywords, which indicate the size of your niche

>   Highlight points for consideration

>   Duration / Cost to App development

>   How you can start the Apps development straight away

If you are looking for private consultations either for yourself, or for your company, then suggestion three is for you...

Suggestion 3
If you like the private experience either for yourself personally or your company then this is the option for you...

The fast-track program. We will assess where you are, and take you forward giving you all the support and guidance you need at every stage of the journey. However we only like to work with those that have already attended our courses, watched our DVDs or read this book, because we know those are the people who will be most successful.

So please send an email requesting the *fast-track course*, giving the barcode number from this book as proof to *info@thinkemotion.co.uk* and we will introduce you to our *fast-track* program.

Which suggestion is right for you?

**www.AppManSecrets.com/
RichAppPoorApp.html**

BONUS:
iPhone sketch templates

Fancy sketching out your own App?

Simply sketch away using these outlines as a guide. There are several pages here to get you started. Alternatively you can download these templates along with other book bonus material from *www.AppManSecrets.com/RichAppPoorApp.html*.

Have fun!

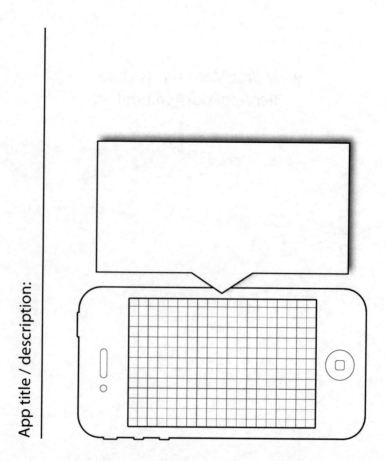

App title / description:

BONUS:
iPhone sketch templates

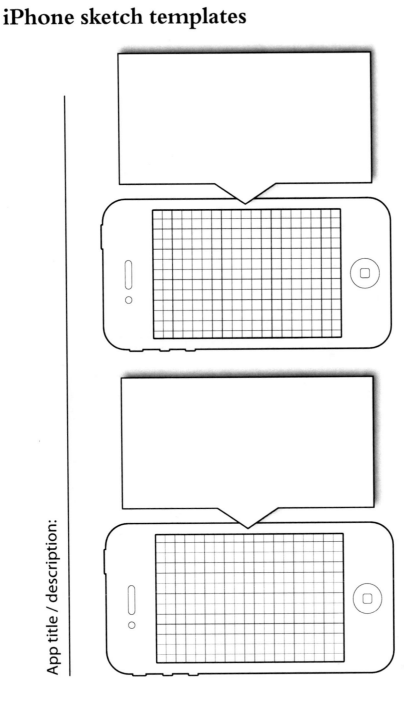

App title / description:

BONUS:
iPhone sketch templates

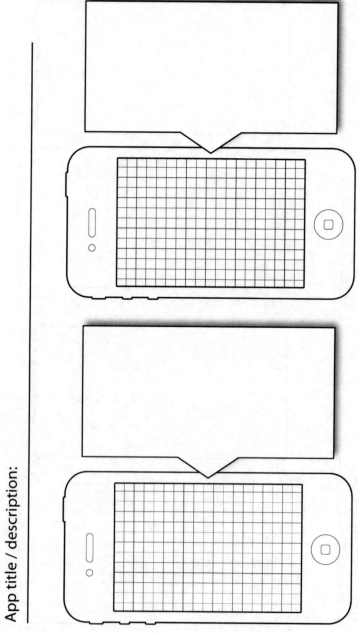

App title / description:

BONUS:
iPhone sketch templates

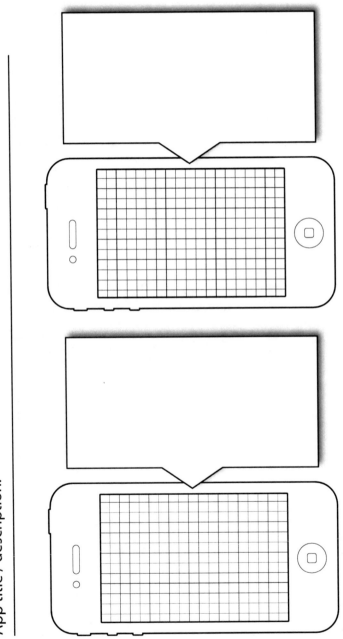

App title / description:

BONUS:
iPhone sketch templates

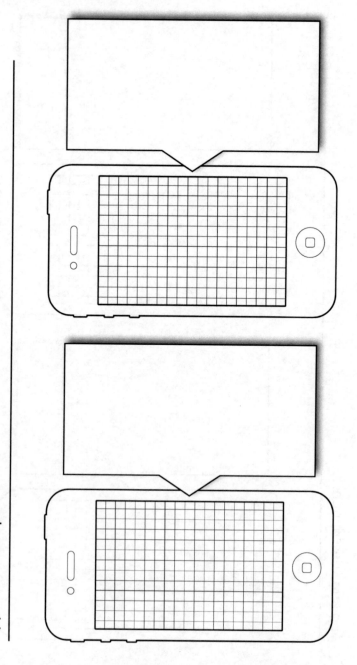

App title / description:

BONUS:
iPhone sketch templates

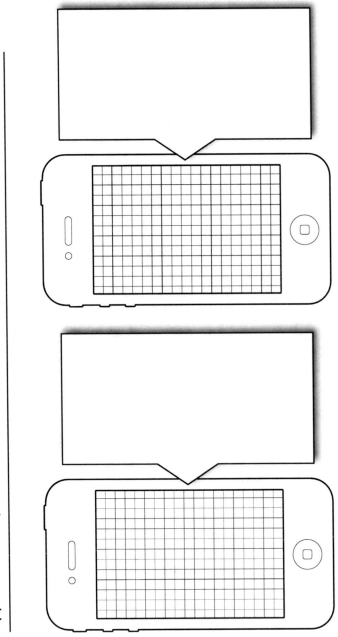

App title / description:

BONUS:
iPhone sketch templates

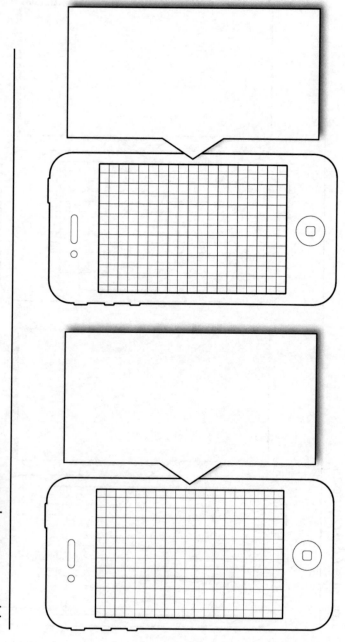

App title / description:

BONUS:
iPhone sketch templates

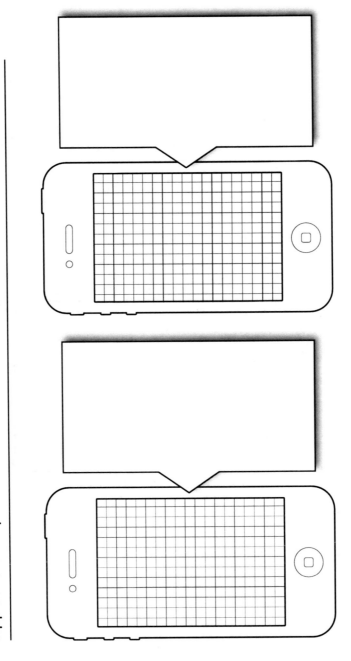

App title / description:

BONUS:
iPhone sketch templates

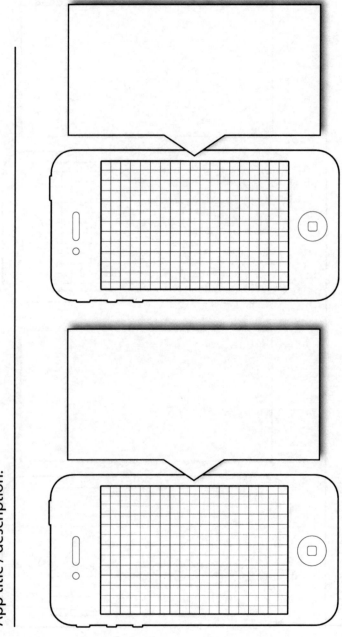

App title / description:

BONUS:
iPhone sketch templates

App title / description:

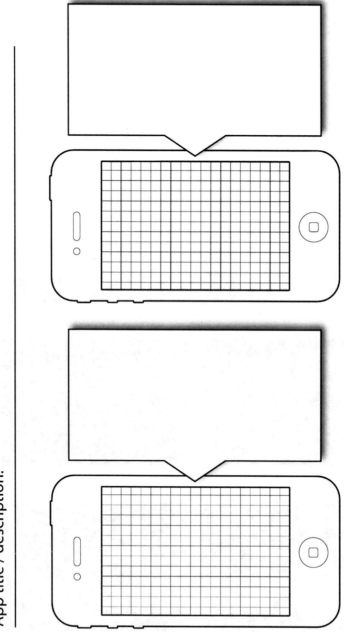

BONUS:
iPhone sketch templates

Fancy sketching out your own App?

Using a marker pad (or tracing paper) which you can buy from any good graphic suppliers, simply trace the outlines from the previous pages, and then overlay these elements and sketch as required. *Give it a go, it's fun.*

BONUS:
iPhone sketch templates

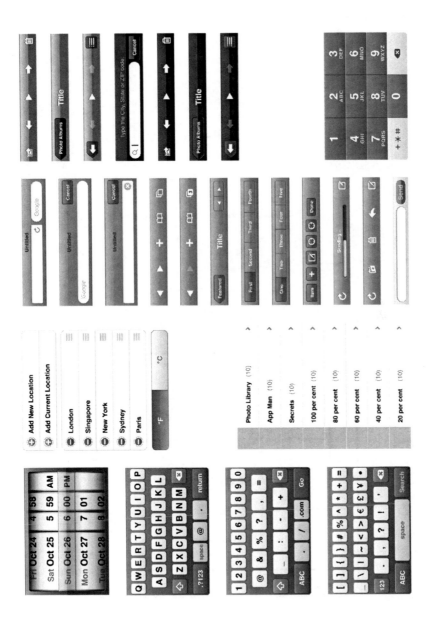

BONUS:
iPad sketch templates

Fancy sketching out your own iPad App?

Simply sketch away using these outlines as a guide. There are several pages here to get you started. Alternatively you can download these templates along with other book bonus material from *www.AppManSecrets.com/RichAppPoorApp.html*.

Have fun!

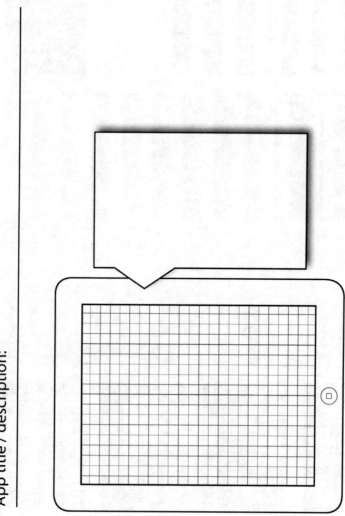

App title / description:

BONUS:
iPad sketch templates

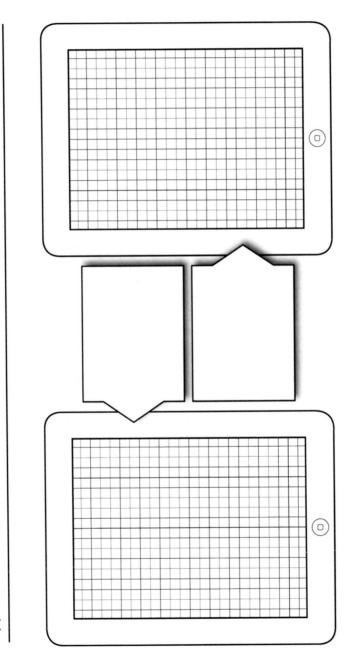

App title / description:

BONUS:
iPad sketch templates

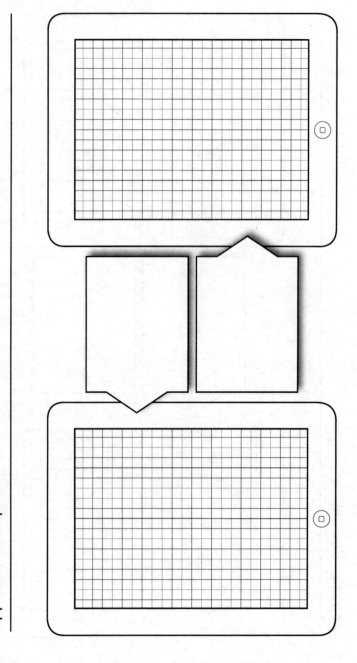

App title / description:

BONUS:
iPad sketch templates

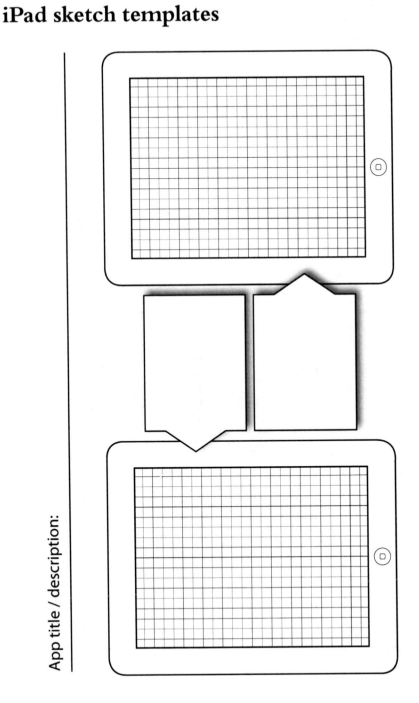

App title / description:

#RICHAPP

BONUS:
iPad sketch templates

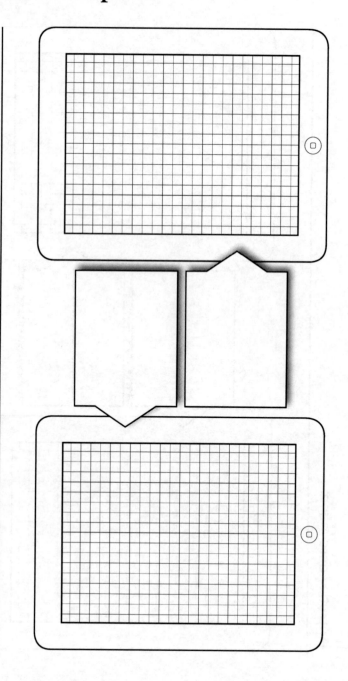

App title / description:

BONUS:
iPad sketch templates

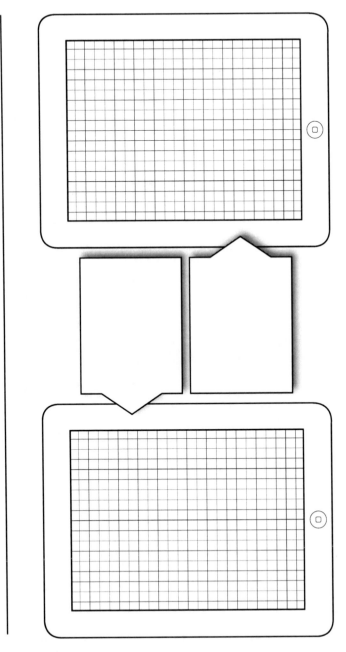

App title / description:

BONUS:
iPad sketch templates

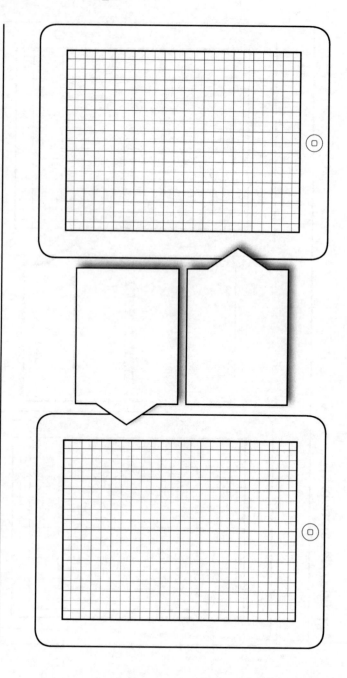

App title / description:

BONUS:
iPad sketch templates

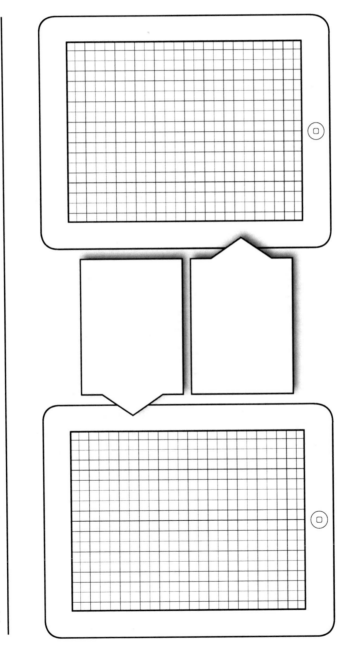

App title / description:

Index

Index

Connect

You can connect direct to the Author, Simon K Williams using these links;

Facebook:
Simon Williams The AppMan

Twitter:
Simon_ TheAppMan
#RichApp

Lightning Source UK Ltd.
Milton Keynes UK
UKOW052316020413

208534UK00001B/1/P